Clearing Spaces

CLEARING SPACES

INSPIRATIONAL TECHNIQUES TO HEAL YOUR HOME

BY KHI ARMAND

STERLING ETHOS
New York

An Imprint of Sterling Publishing Co., Inc.
1166 Avenue of the Americas
New York, NY 10036

ISBN 978-1-4549-1958-2

Library of Congress Cataloging-in-Publication Data
Names: Armand, Khi, author.
Title: Clearing spaces : inspirational techniques to heal your home / Khi Armand.
Description: New York, NY : Sterling Ethos, 2017.
Identifiers: LCCN 2016030673 | ISBN 9781454919582 (paperback)
Subjects: LCSH: Housing and health. | Sick building syndrome. | Mind and body. | BISAC: BODY, MIND & SPIRIT / General. | HOUSE & HOME / Cleaning & Caretaking. | BODY, MIND & SPIRIT / Spirituality / Shamanism.
Classification: LCC RA770.5 .A76 2017 | DDC 613/.5--dc23 LC record available at https://lccn.loc.gov/2016030673

Distributed in Canada by Sterling Publishing Co., Inc.
c/o Canadian Manda Group, 664 Annette Street
Toronto, Ontario, Canada M6S 2C8
Distributed in the United Kingdom by GMC Distribution Services
Castle Place, 166 High Street, Lewes, East Sussex, England BN7 1XU
Distributed in Australia by NewSouth Books
45 Beach Street, Coogee, NSW 2034, Australia

For information about custom editions, special sales, and premium and corporate purchases, please contact Sterling Special Sales at 800-805-5489 or specialsales@sterlingpublishing.com.

Manufactured in China

2 4 6 8 10 9 7 5 3 1

www.sterlingpublishing.com

To Earth's wisdom-holders,
and to the ancestors
who have yet to come

CONTENTS

INTRODUCTION

The history of energetically clearing spaces is probably as old as the earliest human-made dwellings. Rather than being complex ritual acts, these practices were most likely quite similar to ones found in traditional folk magic around the world today—an extension of overall space-cleaning efforts. As early humans interacted with both seen and unseen elements in their environment, they discovered that the natural world and its governing forces had their own stories to tell regarding which areas would be welcoming to humans and which were off-limits. Spaces might even be marked as hallowed ground; in a broad natural environment, some spaces were considered to be of even greater sacred significance. It was, therefore, natural that such an idea would extend to the spaces in which humans actually dwelled both collectively and in family units. Just as nature and the great forces it held needed to be respected, so was there a need to protect from intruding elements—non-human predators both animal and spirit. And just as nature provided medicine for healing and protecting the body from invasion, so did it provide cures for human habitats as well.

It was in college that my own foray into space clearing began. Having been granted a private dorm room in a larger, more chaotic dwelling shared with students from different backgrounds and having different needs, it became clear to me early on that greater delineation of my space beyond the walls that bordered it would be necessary if I were to achieve my academic goals. Not being the most organized person at the time, I also noticed that my focus and productivity suffered if I did not tend to the messiness of my own creation by my accumulation of books, papers, dirty clothes, and dishes. Eventually, with the stress of school responsibilities and the high-flying emotions of being college age, even this wasn't enough. It was during this time that I began my first clearing rituals, smudging my room weekly with burning sage, followed by sprinkling with saltwater around the perimeter. In this, the elements of Fire, Air, Earth, and Water were all present, and the addition of the popular Nag Champa incense "for good vibes" settled my intentions for my environment.

In truth, this wasn't the first time I'd used those elements to cleanse a space; as a solitary Neo-Wiccan practitioner, I'd use them whenever I was conducting a ceremony indoors after "casting a circle" with my ritual wand to delineate sacred space for my tasks. My smudging and aspersing would be performed in a *widdershins* (counterclockwise) fashion to banish unwanted energies, while acts of blessing with incense would take place while walking *deosil* (clockwise) around the inside of my circle. What changed was that I began to see my living space as hallowed ground and the magic of my mundane life deserving of the same level of care and attentiveness that I brought to my religious one. When more than a week would pass without these simple acts, I found my mental health and productivity suffering. So my journey into serious study of the ins and outs of space clearing began.

This book is a culmination of methods derived from my experience as a professional shamanic practitioner and hoodoo root doctor within the Southern Conjure tradition of American folk magic. Rather than being a primer for the spiritual hygiene of a place, it is meant to equip both novices and advanced spiritual practitioners with a deep and practical understanding of environments from an energetic perspective, along with tools and techniques to tackle energetic issues in a space for long-term resolution. Serious emphasis is also given to investigative methods, such as divination and shamanic journeying, to help practitioners arrive at customized prescriptive acts directly applicable to the issues they encounter for maximum precision and efficacy.

CHAPTER 1 of this book explores the core tenets of animistic cosmologies inherent in shamanic and folk magic practices. In it we explore the doctrine of signatures, plant spirit consciousness, and commonalities in folk magic practices around the world.

CHAPTER 2 looks at the materials and methods that will be a focus in this book, ranging from *curios* found in the African-American rootwork tradition to other tools practitioners may find useful. This chapter offers a lens into how spaces interact with the living, encouraging creative responses to issues a space may have.

CHAPTER 3 explores acts of cleansing and clearing, from floor washes to smudging, for removal of energetic debris and renewal of a space.

CHAPTER 4 takes a look at protection for the home, including investigating why some homes naturally are more protected from intrusive energies than others. It provides tools for warding and guarding spaces according to the needs of the environment.

CHAPTER 5 explores the issues of hauntings and intrusive sentient entities and how they can be addressed and protected against through acts of exorcism and by working with materia magica best suited to these concerns.

CHAPTER 6 deals with spirits of place—from the sentient energies associated with a home or building to land and nature spirits beneficial for practitioners to be in relationship with, as well as identifying ones to avoid. This is one of the least-explored topics in space resolution; this chapter discusses in depth how relationship with these beings is perceived in cultures around the world, thereby offering a deeper understanding of avenues to consider taking when issues arise.

CHAPTER 7 looks at various tools of divination and ways they can be used to accurately diagnose energetic disturbances and point the way toward effective acts of remediation.

CHAPTER 8 explores possibilities from around the globe for working with saints, angels, and other helping spirits to help maintain protection, peace, and prosperity in an environment.

It is my hope that this book will equip you to engage with your environment more deeply, make allies amongst the land spirits and inhabitants of the green world, and craft spaces that support both yourself and others.

CHAPTER 1

Diagnosing and Remediating Spaces

The ability to diagnose the energetic health of an environment is as instinctual in humans as it is in other animals. This inborn perception is experienced across a range of a person's consciousness, depending on the types of energies that are encountered. The most obvious of these perceptive glimpses are the "gut instinct" and "fight-or-flight" responses we have when we sense imminent danger in a space. On a biological level, we know that our lives are threatened and we feel the urge to flee. A level deeper than this, perhaps, is what we experience when a space is under construction or we are standing on land that once was bountiful but now is barren. Though neither situation poses a direct threat to life, neither is conducive to the growth and expansiveness that living entails. We shrink back, knowing that settling into these kinds of environments long-term would be an exercise in futility.

The next level of experience of a space happens where our bodies and the world of our emotions meet. It is here that we might have the experience that "something just doesn't feel right," be it a sense of awkwardness whose origins are difficult to discern or a general feeling that a room isn't aligned with our aesthetic expectations. It is this experience that leads people to rearrange furniture, purchase wall hangings, or bring in potted plants, as our personal sense of well-being is directly related to the health of our environment, and this includes how textures, colors, woods, metals, and other aspects of décor interact with one another.

The lengths we will go to get a room "just right," even in cities like New York, where the average apartment renter moves every few years, are pretty justifiable. Whether in the living room, an office, or a corridor, human beings instinctually expect a level of comfort from their environment, because comfort is the reason indoor spaces exist in the first place. When something in a space is at odds with our comfort level, we instinctively make changes to the space, often starting with an assessment of clutter or lack of physical cleanliness and remediating problems of that type.

There is a level of diagnosis more abstract than any of these, however, that explores whether or not a space is conducive to its function. In addition to being comfortable, does it support the activities that one intends to carry out within it? More than a kitchen requiring a stove and a bedroom a bed, does the former space tend to make cooking tedious or a joy? Does the latter environment make it easy to sleep, or is it arranged in a way that makes sleep nearly impossible? At this tier, personal taste in regard to aesthetics matters most, as do social and cultural norms for how a space should be laid out and what should be in it.

In every traditional culture on earth, it is understood that there is an energetic plane that exists alongside our physical one, and that what we respond to mentally, emotionally, and behaviorally in a space isn't always obvious to the physical eye. The goal of remediating the causes of discomfort and dis-ease in an environment is rarely just bare-minimum survivability, but is holistic, seeking to make the space conducive to the total range of needs and goals of its occupants. Hence, in the East Asian school of feng shui, there are traditional bedroom cures and taboos if one is single and seeking a mate, specifically regarding the placement of the bed. Such forms of remediation manipulate the physical world to shift the energy of an environment, similar to the way the physical cleanliness of a space makes it more conducive to mental focus and peace of mind.

Other acts of environmental remediation in traditional cultures specifically target the energetic plane, whether to extract undesirable energies or enhance certain properties in a space (such as its conduciveness to romance). Physical tools such as plant matter, stones and minerals, incense, candles, and liquids like water and alcohol might be used for their energetic or symbolic properties as tools to create change, but they are not the focus of the remediation itself. Similarly, the use of external tools and inner experiences, such as sound and trance that shift one's consciousness toward a clearer perception of the energy of an environment, are meant to affect the space on the subtle plane only. It is these tools, and how they are used in American folk magic and neo-shamanic practices, that are focus of this book.

Folk magic refers to a broad variety of cultural practices around the globe that are meant to affect spaces, persons, and events for the purposes of protecting, healing, removing intrusive energies, attracting that which is desired, and innumerable other intentions. It is fair to consider it related to other forms of "magic" found in the popular cultural imagination, but what differentiates folk magic is its common usage, to varying degrees, throughout and within a specific cultural group. Practices within folk magic—which might additionally be categorized as "folk healing"—are relatively simple and tend to require minimal tools and ingredients, and many of these are often easily available as common household items. Techniques are often passed down within families and are often so woven into day-to-day home life that they are not seen as particularly "magical" at all but simply actions toward necessary or desired results.

American folk magic is a conglomeration of sacred practices found throughout the cultural groups that were brought to and that immigrated to the United States over the past four hundred years. Many of these indigenous traditions became syncretized with Protestant and Catholic forms of Christianity and further synthesized with one another, as is the case with African-American conjure, or hoodoo. Originating in the southern U.S. as a mixture of Congolese sorcery techniques, American Indian herbalism, and Protestant Christian theology, hoodoo is famous around the world today for its practicality, efficacy, and richness of lore, the latter immortalized in numerous blues songs recorded during the mid-twentieth century. By then, the practice of hoodoo had incorporated Jewish Kabbalistic symbols and Dutch-Germanic powwow practices, embracing them toward the creation of a definitively American magical cosmology.

"Shamanism" is a term widely used in the world today that refers to an array of techniques for contacting and working with helping spirits in the spirit world, especially for the purposes of healing. The term is rooted in the Siberian Evenki word "saman," with a root prefix meaning "to know." Western cultural anthropologists studying North Asian tribal roles during the twentieth century began applying the term to similar indigenous spiritual practices found across the world. One of those anthropologists, Michael Harner, became known for bringing what he saw as the most common principles found throughout these traditions to the U.S. and creating a body of practices from them, popularly known as "core shamanism," through the Foundation for Shamanic Studies. The technique known as "shamanic journeying," the act of traveling the realms of the spirit world using sound as a vehicle, is the most well known of these practices. Though the drum remains the most common instrument for such purposes and is widely considered to be "traditional" in the West due to the indigenous cultural practices it is featured in around the world, journeying can effectively be done with other instruments, as well as with prerecorded instrumental and vocal tracks. Techniques for effectively journeying to diagnose and resolve energies in physical environments will be explored in this book.

ANIMISM

At the root of the world's shamanic traditions and the folk magic practices born from them is an animistic worldview that posits that the earth is full of beings,

both human and non-human, that all deserve our respect and awe, and that sometimes the non-human ones can be allied with in order to effect change and help ensure well-being. From the steppes of Siberia to the rainforests of the Amazon, plants, minerals, rivers, lakes, and even the larger ecosystems that they are a part of are seen by traditional peoples as sentient beings with feelings, thoughts, and power in alignment with their form and the role they play in their ecosystem. They have opinions about how they are treated and perspectives on both human evolution and the greater cosmos. Though it is the responsibility of everyone in a culture to adhere to traditions and norms that ensure right relationship is maintained with these surrounding and sustaining environmental forces, it is the job of those chosen by spirit or by family lineage to take up a shamanic role and to propitiate and appease these forces—from gods to nature spirits—when balance is not maintained. Such is the reverence that is held for these forces that ensure rain for crops, crops for sustenance, herbs for medicine, fire for warmth, animals for hunting, and all other basic needs.

The inherent consciousness of plants is not to be overlooked in this context, and it is the special reverence diverse cultures have held for certain types of local plants that has led to the smudging rites that have remained popular even today. Though the indigenous technique of *smudging*—or burning dried plants to help clear harmful energies from a person or place—has become well-known even in relatively secular industrial Western cultures, the plants considered most sacred for use in this rite vary by region. Amongst the Lakota, white sage is called "Grandfather" and is the most popular plant called upon for this task, while among practitioners of Northern European shamanic traditions, mugwort fills that role. Among the Tuvan shamans of Siberia, juniper is one of the most revered plants for performing this work, and it is said to contain medicine in its burning smoke that helps ensure harmony and well-being in a person it is wafted over.

PLANT SPIRIT CONSCIOUSNESS AND THE DOCTRINE OF SIGNATURES

The plants that are most revered by a culture for their cleansing and curing properties because of a deeply sacred bond between the people and the plants' spirits are not the only ones that are sacred, however. From aiding the healing of physical ailments to psychologically empowering warriors and hunters, plants play a role in nearly every aspect of traditional culture, and their efficacy in resolving physical issues sometimes mirrors their ability to effect changes in non-ordinary reality. Across the globe, stories tell of plant spirits coming to people residing close by in visions and dreams to instruct humans in proper plant use. Using the doctrine of signatures, humanity has worked effectively with these spirit elders for nearly as long as we've existed.

According to the doctrine of signatures, every plant has physical characteristics that correspond to the virtues that make it helpful for use by humans. The concept, written about by sixteenth-century Swiss-German philosopher Paracelsus, exists in various cultures around the world and was known even in ancient Rome by philosopher Dioscorides, who described signature plants as far back as 65 CE. The plant eyebright, with blossoms that mimic the human eye, has long been used in European magical traditions to treat optical diseases, as well as issues of mental and spiritual clarity. In South American Amazonian traditions, the leaves of the *lengua del perro* ("dog's tongue") plant are employed to promote loyalty in romantic relationships, mirroring the generally amicable symbiotic relationship humans

have had with canines throughout the millennia.

It is the combination of lore surrounding a culture's most sacred plant allies, certain people's gifts within that culture for communicating with spirit citizens of the green world, and knowledge of practical application of plants that gave rise to what we know today as traditions of folk healing, folk magic, and folk curing—familial and specialized ethnobotanical practices that have endured despite the loss of shamanic ceremonial roles and traditions. These practices range from the crafting of poultices, baths, washes, oils, and fumigations to the more overtly occult creation of charms, talismans, powders, and potions. It is important to note that it is only within the Western philosophical framework of rational dualism that views the physical body as separate from the mind and emotions that a distinction such as this would be made; within animistic folk and shamanic cosmologies, all forms of "medicine" are deemed spiritual, with Spirit the common denominator of all that is.

Working with botanicals and other *materia magica* spiritually means more than simply adding them in the correct amounts to potions and incense. Though every practitioner has a unique way of working, the minimum I suggest is an expression of gratitude to the materials being worked with, followed by an expression of intention to ally with the goal at hand. Recognizing the sentience of the materials that will become materia magica by use, and communicating with them, is at the heart of animism. Like trusted friends, the leaves, stones, and bones are asked to bring their perspectives to bear about the issue at hand.

Though significant differences are found in the symbolism used throughout the world's folk healing traditions, similar practical applications of plants and other materials, such as minerals, liquids, and soil types, are found across cultures, and the materials are used in healing based on their color, personality, common usage, proximity to places of power, and additional cultural lore. No matter the context, all are seen as sentient and powerful. When masterfully employed, alone or in combination with other materia magica, they have the ability to effect changes in reality.

The materia magica of African-American "rootwork" will be explored in the next chapter, as well as the core trance technique of neo-shamanism known as journeying.

CHAPTER 2

MATERIA MAGICA

In this chapter, we'll touch upon the basic tools and modalities that will be explored throughout the rest of this book.

VISUALIZATION

Visualization is consciously choosing to have an inner experience of something without it necessarily occurring physically or in the present moment. It involves the use of one's imagination. Though it is often described as using one's "inner eyes," the richest and most effective visualizations involve the use of all of one's inner senses. Visualization is a powerful tool in magic, as it can create an experience in the present moment of an event or circumstance that you want to attract or achieve in the future. From executive boardrooms to the realm of professional sports, visualization is considered a useful tool everywhere.

Being able to effectively visualize on demand takes practice, but once you learn how to do it, whole worlds can open up to you. Here's an easy exercise for practicing your visualization skills:

Close your eyes and imagine that you are holding an apple. Using all of your senses, explore the fruit. Does it feel waxy or smooth? What color is it? Bite into the apple. Feel it crunch between your teeth, and hear the sound. Does the apple taste sweet or sour? Can you smell the rich aroma of the fruit? Visualize the apple disappearing, bite after bite, until it is just a core. Now, visualize it as whole again. What's different about it this time?

Visualization can be used to empower magic spells and as a stand-alone tool to shift experiences through intention, as we'll be exploring in the next chapter.

WHAT MAKES MAGIC WORK?

Theories abound as to why some people find what we consider "magic" effective. Aleister Crowley, the infamous occultist and founder of Thelema, called magic "the Science and Art of causing Change to occur in conformity with Will." Some theories put special emphasis on the mind and personal willpower of the practitioner, while others especially emphasize the importance of the materials used in the making of magic spells, including the popular notion of "the rarer, the better." And some believe that a special power is held by those who are "gifted" to effect change via spiritual means by virtue of the conditions of their birth, such as their place in the family constellation, being the bearer of a magical family heritage, and/or being born with certain physical traits that denote magical ability. There are other camps that swear that certain magical traditions are more powerful (and, therefore, more dangerous) than others, and that any work done with those traditions is bound to be more effective by virtue of the secret technologies wielded by savvy practitioners. This latter category is almost always shaped by the notions of exoticism that exist in every culture, since magic, by its very nature, engages the mysterious and the unknown.

In my own experience, it is a combination of factors that can help any discerning practitioner to create effective change through magical means:

PERSONAL POWER: Every magical system emphasizes the power of the practitioner as a part of creating effective magic. Though some of

this power may be believed to be endowed by non-human forces, even those who are considered gifted must hone and maintain their gifts and vitality. Practices that reinforce the practitioner's connection to nature are often considered important, as are actions that generally encourage wellness of body, mind, and spirit. With all magic being, in some ways, a moving into the world of the practitioner's will, the ancient adage "Know thyself," found inscribed at the famed oracular site at Delphi, Greece, is widely applicable. Self-knowledge entails, in the short-term, a willingness to explore one's deeper motives and, preferably, a lifelong dedication to self-exploration.

Personal power can also be enhanced through spiritual means, such as the creation of talismans crafted for such an intention, the engagement of allies found in the natural and spiritual realms to help augment one's own ability to have an influence, and practices such as prayer and meditation, which help center and align the practitioner with forces that help ensure vitality.

FOCUS AND SPECIFICITY: A strong, clear intention is the basis of almost all powerful magic, and that specificity is reinforced through the use of materials that are most in alignment with the practitioner's intention. For instance, a spell to obtain a job should not be mistaken for a spell to "attain wealth," and if there are specifics that are desirable relative to the employment that is sought, those should be included. General intentions might attain general results—or none at all.

APPROPRIATE TECH: From candles to mojo bags, baths and washes to spirit pots, there are an incredible number of technologies for achieving desired results in magic, with most cultures having similar magical technologies, in spite of distance. From ritual bathing and washing in herbal infusions to the carrying of talismans and medicine bags, magic across the world is far more similar than most imagine. What matters most in choosing among these technologies is knowing which ones are most appropriate to the situation at hand.

SYMPATHY AND CONTAGION: As a rule, the concepts of sympathy and contagion matter greatly in nearly all magical traditions for increasing the chances of achieving one's desired goals. Contagious magic includes the direct application of magical technologies onto a person, place, or thing and is often considered the most powerful. Ritually washing a home with choice botanicals to help ensure its protection is generally considered more powerful than lighting a candle from afar for the same intention, though the latter might indeed be both effective and the only reasonable course of action in certain circumstances. Sympathetically, in lieu of being able to apply materia magica to a space directly, a practitioner might opt to create a representation of the space and perform magical work on it to achieve the same effect.

MEANINGFUL AND CONGRUENT SYMBOLISM: Our world is full of diverse spiritual traditions that have arisen from distinct cultures whose ways of life have been tied to distinct regions and climates. Though cultures may work with similar magical technologies, the symbols and spirits employed and invoked tend to differ greatly. When crafting magic, it is important to choose symbols and energies that resonate both with you and with one another.

HOODOO FOLK MAGIC

When enslaved Africans were brought to the New World, they brought along with them the medicinal and magical traditions of their homelands, from Sierra Leone to the Congo. In the New World, these traditions were stifled and banned, but through the courage and cleverness of the people who brought them, a large number of the practices and lore survived. In Catholic Cuba, the orisha spirits of the Yoruba religion were venerated in the guises of Catholic saints, and a similar process of masking occurred in Haiti and Brazil. In the Protestant U.S. during the seventeenth and eighteenth centuries, however, with little Catholicism existing in the American colonies and, therefore, no saints to hide the spirits behind, what remained was primarily Congolese magical technology, with traditional symbolism and religion replaced by the Bible and Christianity. In the rural South, this unique blend of African spiritual techniques, Christian texts, and the Judeo-Christian view of the deity became what is known today as hoodoo, conjure, or rootwork.

From the 1700s 'til today, rootwork has been an important wellness modality amongst Black Americans, with plants, minerals, animalia, and common household items used to treat everything from earaches to issues of love or money. If someone is experiencing ill health, financial difficulties, or other losses, the person is said to be under *crossed conditions*, and rituals prescribed by a professional root doctor, or that are known by someone in the person's family, can help to remediate these conditions and invite well-being into the person's life.

What began as a rural, community-based practice of folk medicine and magic became known nationally through the mass production and direct-mail advertising of hoodoo curios in the mid-twentieth century, turning beloved hoodoo remedies that had been passed down for generations into mostly inert condition products consisting of powder and oil bases to which fragrance had been added. But at the same time, artisanal-quality oils, baths, washes, and powders with genuine plant, mineral, and animal material continued to be crafted in small batches by those who had not forgotten the old ways.

Some of the tools and techniques used widely in hoodoo include:

CONDITION PRODUCTS: Botanicals, minerals, animalia, and essential oils may be added to oils, powders, bath crystals, alcohols, and other menstruums and bases before being prayed over and empowered to aid practitioners in achieving desired outcomes. Oils may be used to dress or anoint a person, place, or object to be imbued, and powder can be similarly dusted over papers, sprinkled on the ground to be walked over, or blown into a room. Baths, which may simply be ready-made combinations of botanicals or minerals for steeping as an infusion, can be taken to *uncross* conditions or attract desired experiences, and these items can similarly be added to detergents to make a magical floor wash or added to spray bottles for applying intention to a space. Though there are a few formulas that are considered "traditional," based on how long they've been in circulation, most hoodoo root doctors pride themselves on crafting powerful recipes for use in their own practices.

WRITINGS AND SYMBOLS: Petitions and intentions may be written on pieces of paper to be put beneath candles, added to talismans, placed in one's shoe, or used in a variety of other ways. Symbols like the Christian cross, hearts, or eyes may be drawn, and portions of biblical scripture applicable to the intention may be written out as well.

SOUND AND BREATH: Prayers may be recited over magical work, persons, or places, including, but not limited to, the recitation of psalms and other biblical texts. Items like dolls and mojo bags may be given special names and consecrated "in the name of the Father, the Son, and the Holy Ghost" after being breathed to life to work as conscious entities.

CANDLES AND LAMPS: Oil lamps and candles of different shapes and sizes whose colors correspond to different conditions may be *fixed* and burned to help achieve an intention—*fixing* is adding condition oils, botanicals, or minerals corresponding to the outcome desired.

Though you can purchase quality condition products from reputable retailers, you may wish to experiment making your own. The following are some condition oil recipes from my own formulary. They are based in hoodoo correspondences and can be blended for use in clearing spaces. When available, a few drops of essential oil can be substituted for dry botanicals, but fresh botanicals should never be used; they are hard to grind into powder and will make oils rancid.

Practitioners who craft their own magical oil tend to have one favored base oil that they use for all of their blends, or they choose amongst the many available ones based on further magical associations. Olive oil is a very excellent standard base for use in making oils within the hoodoo tradition because it is commonly available, as well as for its association with biblical texts. Still, others might prefer to use sunflower seed oil when crafting blends for benefic purposes and use other bases as desired. To extend the shelf life of these oils, I suggest adding a dropperful of jojoba or vitamin E oil per half ounce, and I suggest making no more than half an ounce of an oil while you experiment with amounts of material and scent (if that is important to you). I tend to make my oils with fractionated coconut oil because of its total lack of scent and the fact that it never goes rancid. For best storage, keep in tight-lidded glass bottles out of the sun.

UNCROSSING: Lemongrass, hyssop, devil's shoestrings (cramp bark), salt

VAN VAN (for cleansing and turning bad luck into good): Lemongrass, a small piece of pyrite

PROTECTION: Basil, angelica root, ginger, salt

BLESSING: Angelica root, cinnamon, frankincense, benzoin resin

HEALING: Essential oil of bergamot orange, lemongrass, wintergreen

ALL GOOD WORKS OIL: The simplest way I know of to prepare an oil that can be used to enhance all benefic magic, including blessing and protection, is to pray Psalm 23 over a small amount of olive oil, making sure that your breath and words meet the oil.

SHAMANIC JOURNEYING

Shamanic journeying is a technique known by many names in indigenous cultures around the world and has been gaining popularity in the U.S. over the last forty years. It is a trance state induced through the vehicle of sound, in which the journeyer traverses the spirit world with the aid of spirit helpers to receive information, diagnose conditions, resolve energies, retrieve medicine, or simply interact with helping spirits and other entities. Though the physical body stays in the physical world, an intentional shift in consciousness allows the practitioner to enter non-ordinary reality. In the West, it is primarily used as a tool by practitioners of "core shamanism," a body of practices and techniques taught by anthropologist Michael Harner, creator of the Foundation for Shamanic Studies.

Shamanic journeying is a simple enough practice to learn, but it can be difficult to learn to do well. Much can be accomplished by the novice, but a mastery of the technique requires practice, intimacy with one's own helping spirits, the ability to ask effective questions, and the ability to interpret answers given in journey with an eye toward taking concrete action in the mundane world. Mastering shamanic journeying is beyond the scope of this book, but interested parties would do well to investigate the resources in the back of this book.

The emphasis on developing relationships with helping spirits is not to be overlooked, as the spirit world is too vast to be effectively and safely explored on our own. While journeying, we present our questions to our helping spirits for

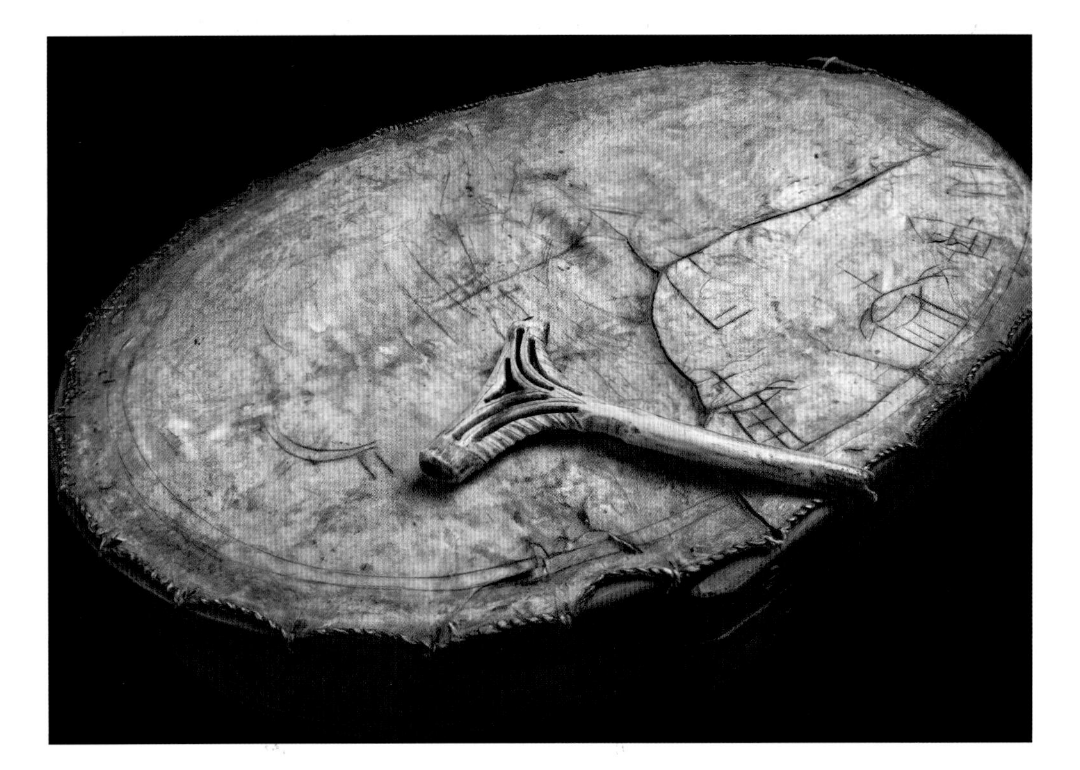

them to answer, are led by them to other spirits and places, can receive their aid in dealing with intrusive energies, and can even embody them to augment our power with theirs when handling especially difficult energies that we may encounter. Many begin the practice by taking journeys to meet their own helping spirits and form relationships with them, and this is highly advisable.

Once relationships with one or more trusted helping spirits have been established, the practitioner can undertake a journey to investigate the energies of a space with relative safety and efficacy. There are various skills that can support the journeyer in this regard, including cultivating trust, clarity, and discernment while journeying, which the Felt Sense Exercise (explored in chapter 7) can aid.

When I began exploring shamanic journeying, I primarily used the practice for open-ended visits to my helping spirits, resulting in half-hour-long experiences that then took at least ten minutes to write down. Having neither additional time nor the patience to interpret them, I found myself deterred

from the practice. Journeying on one question at a time and considering the whole of the journey experience to be the answer radicalized my process, cut down my journey time immensely, and increased my efficacy in both journeywork and interpretation tremendously.

Journeying can be done with live instrumentals or to prerecorded music. Most common is a repetitive drumbeat, though it can be done with any sound that won't distract the practitioner from the journey itself. Novice journeyers will most likely have trouble drumming for themselves, so using an audio track or a phone app is perfectly acceptable.

The first skill to develop when learning to journey is relaxation. Lying flat on your back with your eyes closed, scan your body from the top of your head downward, sensing for any areas of tension and relaxing them while breathing deeply into your belly. You may imagine inhaling the energies of peace and relaxation and exhaling stress and worry. Be sure to relax your mind and heart as well. Next, visualize in your mind's eye the numbers from thirteen to one, counting backward into an even deeper state of relaxation. Once you've reached the number one, you should feel as if the floor beneath you is pushing upward. This is a deep state of relaxation appropriate for effective journeywork; practice often to achieve it consistently. Eventually, it will be as easy as flipping a switch.

Once you reach this state of deep state of relaxation, and with the sound vehicle playing, visualize an entryway into the spirit world. It could be a ladder that leads to a hole in a tree in your front yard, or steps beneath the surface of a pond at a nearby park. Using all of your inner senses, and counting backward/downward from thirteen to one into the journey, imagine a door at the end of this path. On the other side of it, you will call forth your helping spirits and will present them a single question or intention that will be the basis for the journey.

One of the biggest difficulties novices tend to have with shamanic journeying is that the information received is highly symbolic, seemingly from a place of non-logic. In actuality, the information tends to be so simple that it can be easy to miss. While journeying, the practitioner's job is to receive the answer, trust that what is being experienced is legitimate and real, and engage with what comes up in the journey. It is vital that one neither try to control the journey nor interpret it while it is happening. Non-ordinary reality is just that: helping spirits may eat us, we may find ourselves transformed into other creatures, and generally every nonsensical possible image is on the table if that's what our helping spirits believe is the best way to convey an answer to us using symbols and metaphors they expect us to be familiar with. You may think it is all happening in your head and that you are just making things up, but part of the power of shamanic journeying is that it is a deep enough trance state to help ensure that the answers we get will often surprise us.

When your journey is through, thank your helping spirits and return to the door through which you entered. Make your way up into your body and slowly come out of your relaxation, counting upward from one to thirteen. Record your experience in a journal, writing it out in first person, present tense, and only then should you begin the work of interpretation. Check back over the journey and interpretation later so that you can begin to understand the symbolic language your helping spirits speak to you in.

Although forms of divination like the tarot can help us explore deleterious factors and helpful forms of remediation in a space, the visuals and felt senses provided by shamanic journeying tell a fuller story, and when communication with sentient beings is necessary for achieving one's goals within a space, few modalities compare.

CHAPTER 3

CLEANSING AND CLEARING

"The smoke of sweetgrass is pleasant to the good spirits. They come to the smoke. They are pleased with one who makes this smoke. They will listen to what makes such a one asks. But the bad spirits come also to enjoy the smoke. So sage must be burned to make them sick."

—TAKES THE GUN, *Lakota Belief and Ritual*

Though advanced diagnosis and prognosis toward resolving energies within an environment might not be a widely accepted concept in contemporary Western culture, all around us we routinely see the intention to create harmony within a space by manipulating the energy within it. Sage, commonly in the form of smudge sticks, is by far the most popular tool used by laypersons, spiritual seekers, and knowledgeable practitioners alike to cleanse a space of distracting, harmful, and oppressive energies and is often treated like a cure-all for lack of harmony within an environment. Even in religious traditions that generally spurn the practice of magic, adherents will nevertheless invite their pastor or priest to bless a new dwelling shortly after moving in, even if there are no signs of paranormal activity. Olive oil or holy water will almost undoubtedly be part of the aforementioned ritualist's arsenal for anointing doorways throughout the space, paired with improvised or traditional prayers.

A practitioner of American folk magic is likely to have a wide variety of tools and preferred material magica on hand for remediating crossed conditions within a space and may employ them in a variety of ways, including sprays, fumigations, and floor scrubs, which will be explored in the next section.

AGRIMONY: This uncrossing herb is also one of the first that is turned to when the harm one is experiencing has been sent intentionally by another; it is said to reverse spiritual attacks back to their senders.

AMMONIA: This sterilizing liquid is considered a strong uncrossing agent for people and spaces when minute amounts are added to personal baths and washes for floors. It is generally considered strong enough to strip away both harmful and benefic energies, so it is often used only in the most dire of circumstances.

BLUING: Though less popular today than many decades ago, laundry bluing is a whitening agent and can be found as squares or balls that are added in small quantities to water infusions used to cleanse spaces.

FLORIDA WATER: Originally a commercially prepared toilet water made available during the nineteenth century, like many common household items, it became a spiritual curio associated with cleansing and blessing due to its inclusion of floral and citrus oils. Many different brands can be found today, and some practitioners opt to craft their own for use in personal bathing or sprinkling rites and as an addition to floor washes and scrubs.

FRANKINCENSE: The holy incense of the Catholic church and one of the gifts mentioned in the Bible given by the Wise Men to baby Jesus, this resin is lauded for its high vibrational frequency and is often burned in rites of spirit removal and blessing. In hoodoo, it has the additional association of helping to empower all good works.

HOLY WATER: All over the world, Catholic practitioners of folk magic can be found employing water in rites of cleansing, exorcism, and blessing. The water has been blessed by clergy and is easily obtained from nearby churches. Other practitioners may opt to bless water themselves through the recitation of prayers or the invocation of helping spirits.

HYSSOP: The preeminent cleansing herb in the hoodoo tradition, it is especially associated with forgiveness and removal of sin because of its mention in Psalm 51: "Wash me with hyssop and I shall be clean."

LEMONGRASS: The primary ingredient in both Van Van oil and in Chinese Wash, it is known for turning bad luck into good luck.

NETTLES: Known for their protective, cleansing, and nourishing properties, stinging nettles are often added to

infusions with agrimony as part of washes and baths for removing crossed conditions.

PINE: The source of turpentine, this tree's resin, needles, and distilled oil are all found in traditional formulas for infusions and fumigations for the spiritual cleansing of spaces.

RUE: A plant lauded for its strength in breaking hexes and curses and removing the effects of the evil eye.

SAGE: Beloved for its ability to clear away stagnant energies and refresh an environment.

SALT: As a cleansing and neutralizing agent, salt is most often employed in the form of seas salt or kosher salt (coarse crystals) in spiritual baths and infusions. Practitioners may also perform their own blessing rites over table salt, reciting prayers and invoking helping spirits.

VINEGAR: Both white and apple cider vinegar appear in traditional hoodoo baths and washes for uncrossing persons and spaces.

WORKING WITH FLOORS

Almost unheard of in mainstream America is the emphasis on floors as one of the primary sites in the spiritual cleansing of spaces, but this is the space prioritized in Southern rootwork and the West African traditions from which it is derived.

In many cultures around the world, sweeping the floor holds greater significance than merely ridding a household of dust. As is common in folk magic, this simple act is transformed into a spiritual technology when coupled with intention; sweeping, and doing so directionally, is a way of ridding one's dwelling of harm both seen and unseen. Today, the broom, known as a besom, is a prized ritual tool amongst Neo-Pagan witches and is used for cleansing a space before ritual occurs; but traditionally, the common household broom would be all that was needed to accomplish a multitude of acts.

The ability to rid one's home of unwanted energies with a broom is not limited to sweeping; many homeowners and shopkeepers will testify that placing a broom upside down next to the door when you want a visitor or idler to leave will result in the person making an exit shortly thereafter. The broom's use as a banishing tool can also be seen in how sweeping someone's foot accidentally is traditionally viewed in the Black-American community—a person whose foot has been brushed may be concerned that the act has negatively affected their luck.

NEW HOME BROOM RITUAL

Old American folklore tells us that it is bad luck to bring an old broom into a new home; doing so is inviting old troubles into what should be a new life. To help leave your troubles behind you when making a move to a new location, make peace with the space you've dwelled in, first by sweeping it out with your old broom before leaving it at a crossroads—a place where two streets intersect in perpendicular fashion. On the way to your new home, purchase a new broom for starting fresh, a loaf of bread so that no one ever hungers, and a new box of table salt.

A more thorough cleansing of a space, especially if someone is found to be suffering from crossed conditions, might call for a floor wash—a rite performed monthly or annually by many hoodoo practitioners for maintenance. A common tool used in this work is popularly known as Chinese Wash, due to the blend of Asian grasses that it includes. Lucky Mojo Curio Company, one of the most popular suppliers of Chinese Wash, happens to include broom corn straws in every bottle, yet another testament to the endurance of this common household tool as a catalyst for magical change.

HOW TO PERFORM A FLOOR WASH

Cleansing of one's home or business premises is undertaken to remove harmful energies, jinxes, and curses. It is traditional to perform this work before dawn and before speaking to anyone.

Depending on the size of the premises, pour a teaspoon or a tablespoon of Chinese Wash into a bucket of hot water. Pray over this infusion that any and all harmful energies that linger in the space

CHINESE WASH

Though recipes for Chinese Wash vary, an excellent one can be found in Stephanie Rose Bird's Sticks, Stones, Roots & Bones: Hoodoo, Mojo & Conjuring with Herbs.

6 oz unscented liquid castile soap
3 oz Murphy's oil soap
6 oz distilled water
½ tsp citronella oil
½ tsp lemongrass oil
7 pieces broom corn straw
Pinch of dried lemongrass

Add all ingredients to a 16-ounce bottle. The mixture can be stored indefinitely. Chinese Wash is used in a diluted form. Add a teaspoon (or more, depending on the size of the space) to a pail with your preferred household cleaner before adding a gallon or more of hot water.

be removed. You can add a commercial detergent as well, depending on the needs of your space; in lieu of Chinese Wash, an infusion of lemongrass will suffice. As led by Spirit, you may choose to add an infusion of other herbs with cleansing properties to your floor wash as well.

With a mop, squeegee, or sponge, cleanse the space from the back room to the front or main room. If you are cleaning a space that is multi-floored, start with the top floor and work your way down. As you cleanse, pray and intend that your space be made clean and new. Some find Psalm 23 to be very appropriate for this ritual. If the space is carpeted, lightly wetting a broom and sweeping it over the carpet in the same fashion works just as well. Make sure to give the entryways of the home extra attention.

RECIPES FOR FLOOR WASHES AND BATH CRYSTALS

A couple of down-home recipes for hoodoo floor washes can be found in Aura Laforest's Hoodoo Spiritual Baths: Cleansing Conjure with Washes and Waters:

HOME CLEANSING FLOOR WASH 1

This floor wash is both physically and spiritually cleansing.

¼ cup Pine-Sol
1 capful Chinese Wash (see page 25)
13 drops pine oil or turpentine

Add all ingredients to a bucket of hot water. Wash from back to front, praying Psalm 23.

HOME CLEANSING FLOOR WASH 2

This floor wash is also good for windows and glass surfaces.

1 cup white vinegar
1 cup lemongrass or pine needle tea
1 pinch kosher salt

Add all ingredients to a bucket of hot water. Pray Psalms 23 and 91 while working.

DIY BATH CRYSTALS

Bath crystals aren't just for personal bathing; they can be added to a bucket of water for cleansing a space. Keep a jar on hand for ease of use.

½ cup of kosher salt or sea salt
½ cup of epsom salt
Essential oils of your choice

Combine all ingredients. Store in a tight-lidded, labeled glass container. Add about a tablespoon of the crystals to about a gallon of hot-to-warm water when you're ready to cleanse your space.

Here are a few recipes from my own personal formulary:

POVERTY AWAY BATH CRYSTALS

Funds running low? This blend clears while inviting prosperous flow.

21 drops of lemongrass
21 drops of spearmint essential oil

Add a few drops of your urine to the scrub water right before mopping.

CLEAR MIND BATH CRYSTALS

Perfect for students, freelancers, and others who seek a calm environment.

13 drops of lemon
7 drops of sage
21 drops of eucalyptus essential oil

These botanicals cleanse and invoke wisdom while discouraging bad habits and emotional stagnancy.

In cases where conditions have been particularly difficult or when it is likely a rite of spiritual cleansing has never been performed within the space, practitioners may choose to move even heavy furniture out of the way to ensure that all areas of a room are addressed. They may, additionally, choose to apply the scrub water to the walls, windows, and ledges within a space, using either a hand-sponge or squeegee, The important thing is to cleanse one room at a time—even if over a period of days—until the task is completed.

When finished, pour the leftover scrub into your front yard or dispose of it at a crossroads. Though apartment-dwellers might express concern about this, as a long-term city dweller, I have found that it is preferable to dispose of the scrub water at a crossroads or into the street rather than flushing it down a toilet, which doesn't ensure that the water has fully left the building.

In our modern age, Chinese Wash or essential oils of botanicals associated with spiritual cleansing can be added to a Swiffer or other automatic cleaner for just as potent an effect, and though protection of a space is explored in greater depth in the next chapter, it is important to keep in mind the Aristotelian notion that "nature abhors a vacuum." When we have emptied a space of that which once occupied it, if we aren't intentional about how we want it refilled, we are simply leaving things up to chance. So after intentionally clearing a space, it is just as important to be intentional about the energies that will fill the area.

Chinese Wash is far from the only commercially available hoodoo product for cleansing homes and other spaces. Reputable manufacturers of time-honored hoodoo bath crystals, herbal baths, oils, and incense can be found online using formulas whose popularity goes back nearly a century. Prayerfully made commercially available products like Uncrossing, 13-Herb Bath, Van Van, Reversing, and Jinx Killer contain genuine botanicals and minerals that remediate the harmful energies attached to persons and spaces and are available as bath crystals to be diluted in hot water, herbal blends to be prepared as tealike infusions, and oils to be added a few drops at a time directly to scrub water. For those who feel inclined to make their own formulas, bear in mind that it may be wise to purchase different brands over time while experimenting with recipe-creation, exploring both the spiritual efficacy as well as the scent.

SPRAYS AND SPRITZERS

Sometimes a space just needs a pick-me-up in between cleansings, clients, or right before an event. Just as inventive practitioners have found ways of ensuring that a bit of spiritual hygiene gets included in nearly every mundane task (from adding some Chinese Wash directly into the bottle of their commercial floor detergent to adding a few drops of Van Van oil to household cleaner spray bottles), so are there ways to insert a bit of magic into any moment.

Many people find the scent of Florida water by itself intoxicating, and practitioners may splash a bit on their outstretched hands, combing downward through their aura for a quick personal cleansing. Likewise, it can be added to a spray bottle or perfume spritzer for use on its own or as a base for additional essential oil blends for quick daily or weekly clearing of a space. To get the full effect, pray over the bottle of Florida

SPACE CLEARING SPRAYS AND SPRITZERS

These blends from my personal formulary get so much use in my household that a bottle of each can be found in nearly every room! Feel free to use them often and abundantly.

UPLIFTING

This recipe is truly burden-lifting and speaks to the clearing power of unconditional love.

8 drops of rose essential oil
8 drops of lemongrass essential oil
10 drops of hyssop essential oil
Commercially prepared Florida water

Add all ingredients to a 2-ounce spritzer or spray bottle and shake well before each use.

UNCROSSING AND PROTECTION

This blend clears and protects in a cinch.

9 drops of eucalyptus essential oil
8 drops of sage essential oil
5 drops of oregano essential oil

Add all ingredients to a 2-ounce spritzer or spray bottle. Top off with distilled water and shake well before each use.

water first, thanking the botanicals within it for their medicine and asking them to be potent in their work for you.

If Florida water isn't your thing, holy water or distilled water can be carriers for essential oils and condition oil blends or a pinch of your favourite bath crystals—the latter is just likely to last longer. Be sure to shake before each use.

SMUDGING, SMOKING, AND FUMIGATING

Most earth-based spiritual traditions recognize and honor the elements—Earth, Air, Fire, and Water—as the most basic building blocks of life, and some take great strides to ensure that each is actively worked with during every rite and ritual to ensure that forces are balanced and harmonized. If floor washes bring the powers of Water and Earth to bear on the crossed conditions of space by way of the medium used and the botanicals and minerals infused within it, smudging, smoking, and fumigating help close the circle by bringing the added powers of Fire and Air, using all four elements as vehicles of the practitioner's will within a space.

Smudging is the popular English word for clearing a space or person with the smoke of burning embers, often sage, and is most associated with American Indian practices. In actuality, the use of burning botanicals to cleanse can be found everywhere, with mugwort the popular choice in Northern European traditions. Smoking a space with incense is found in hoodoo as well, inspired by European influences to the tradition in the early twentieth century.

Though sage is an excellent choice and strongly associated with the spirits of the land in the current day U.S., it is not the cure-all that it is often considered to be. Shamanically, sage has a reputation for "drying out," so it is especially helpful in the clearing away of excess emotional energy caused by stress, depression, and arguments, creating a tabula rasa for the mind and for human activity within a space. It is less helpful in the removal of harmful sentient entities such as ghosts, which is an important consideration when these have been detected as a cause of disease within an environment.

At the time of this writing, the South American botanical *palo santo*—meaning "holy wood"—has become hugely popular in the U.S. and a staple for smudging spaces in many homes and businesses. There is something very Upperworld about this plant; while remaining grounding and centering, it is a potent cleanser and excellent at uplifting the vibration of an environment.

Most of the dried botanicals that can be used in an infusion can be used to smoke a space for either cleansing or protection, but most won't be found in the form of smudge sticks. Some might be available in the form of stick incense, but it is important to be sure that actual botanicals and essential oils were used in their creation. Sprinkling a small amount of dried botanicals or a combination of botanicals and oils onto a lit charcoal in a censer is the go-to choice for most experienced practitioners, but commercially prepared incense blends are easy to find as well. These might be intended to be sprinkled on charcoal or may be sold in self-lighting forms that already have a burning agent—saltpeter—present within them. Be sure to open all drawers and enclosed spaces in the area and ensure that the smoke gets into the corners of the room, where stagnant energies tend to remain hidden.

Here's a personal formula for fumigating a space that has been the target of spiritual harm or energetic threat:

JINX-REVERSING INCENSE

Use this to turn harmful energies back to whoever sent them.

1 part frankincense
2 parts rue
2 parts agrimony

Blend ingredients together. Burn throughout the space on a lit charcoal while verbally demanding that all harm and threats leave the environment and return to where they come from.

HARMFUL OBJECTS

The issue of cursing a home or space and its inhabitants by stealthily placing a "tricked" object somewhere in the environment continues today, though it may not be as common as it once was. Root doctors have long used many methods of divination to pinpoint the presence of such objects for proper disposal before taking further measures to cleanse the space and those affected by the curse. Techniques include cartomancy, geomancy, bone throwing, and sheer intuition, among others. Accounts of rootwork pre-dating the twentieth century even tell of one root doctor who kept an insect in a jar. When it was let loose, it would lead the conjurer to offending objects.

Indeed, if such an object is believed to be at the root of the concerns about a space, it should be found and removed, but when this is not possible, a very thorough smoking of the space is advised, with the intention that the smoke "kill" the "jinxed" object and its efficacy. Smudging rites taking place indoors are often performed with a window or door open for the immediate release of harmful energies, but in this case, it is best if openings remain closed so that the counteracting influences of the smoke can truly settle throughout the space before doors and windows are intentionally opened for the harmful influences to be banished. In the hoodoo tradition, there is perhaps no greater element for clearing a space in this manner than sulfur burned on a charcoal. It is extremely caustic and, like the use of ammonia in a floor wash, is said to clear away everything—both good and bad—so its use should be considered a last resort. All living beings should be made to exit the environment after the sulfur has been set to burn in one or more central places, and none should return until the house has been "smoked out."

If a cursed object is found, it should be buried far away from the environment and covered in salt, a thoroughly neutralizing agent.

Far more common than intentionally "tricked" objects placed within or near a space is the presence of haunted objects—furniture, clothing, and accessories to which spirits of the dead are attached or that are carrying harmful energetic impressions from past owners. These may be passed down within the family or purchased at flea markets or antique shops. The wise shopper will smudge newly bought pre-owned objects shortly after bringing them home, to clear any lingering spiritual attachments. This should be enough if there is not an actual conscious entity attached. If there is, the methods discussed in chapter 5 on haunted spaces and intrusive entities should be effective for resolving those issues.

A friend gives me a ring he found many months earlier that he feels will be a great addition to my toolbox. It is silver-toned with a skull embossed in it—exactly the kind of accessory he knows I would love. As I place it on my finger, I sense an external consciousness interface with my own; it is heavy and erratic. I remove the ring and shuffle my tarot cards. The Four of Swords, showing a man lying down in a church as if dead, comes flying out of the deck. I smudge the ring in the smoke of burning frankincense resin placed on a lit charcoal while saying accompanying prayers to my helping spirits to aid the spirit of the deceased still attached to the ring move on. I bury it in the earth for three days, after which I retrieve the ring—it feels clean and new. I place it on my altar to be consecrated for my own purposes in the near future.

VISUALIZATION AND ENERGY HEALING

Visualization is another helpful tool for clearing spaces, and it should be used in tandem with other forms of clearing. While washing a home, for instance, the practitioner may visualize harmful entities leaving the space, and a white, abiding light restoring it to well-being. Energy healing encompasses modalities like Reiki that—by working directly on the subtlest planes of existence—restore vitality. Those trained in Reiki or similar modalities, or those who simply have a gift for healing, may find themselves able to shift a room's vibration by consciously sending feelings, symbols, and intentions through the hands and/or heart, though it may take time for novices to learn to do so in a healthy manner.

Visualization can also be used as a stand-alone form of clearing in a pinch. In Debra Lynn Katz's *You Are Psychic*, she describes a nifty way of clearing a space of harmful energies by visualizing columns of golden light in each corner of a room that stretch from the ceiling of a space down to the center of the Earth. An additional column is visualized in the center of the room that all of the others connect to.

"Now you can command all the energy in the room that is not in alignment with you and your goals to leave the room through these columns of light. Imagine that the Earth is effortlessly pulling down deep into its center the energy from people, spirits, or other entities; extraneous emotional energy; an energy that is getting in the way of your serenity, happiness, ability to accomplish your tasks, etc. As the energy releases, you can look to see what colors are falling down the columns, or you can just know that they are being released. As you do this, be aware of any physical sensations you may be feeling.

"Once you have grounded the room, it's time to own it. Imagine that you are writing your name, in a color of your choice, across at least four of the walls. See your picture hanging on the wall as well."

Another helpful visualization for clearing a space is imagining a long-stemmed rose. Spin the rose clockwise and send it around the perimeter of the space to collect energetic debris. Drop it to the center of the earth and explode it. Imagine a new rose and spin it counterclockwise and do the same, going

in the opposite direction around the room before dropping it to the center of the Earth.

Banishing rites are another form of visualization that combine gesture, sound, and invocation to accomplish the goal of clearing a space either for general living or in preparation for further spiritual or religious acts. As these fall outside the bounds of folk magic, they are beyond the scope of this book but are deserving of mention due to their efficacy and the fact that they include many of the same elements discussed in this section. For a Protestant Christian rootworker, a parallel might be the audible recitation of biblical scripture within an improvised prayer spoken with great conviction "in the name of Jesus," "by the power of the Almighty God," and/or "by the blood of the Lamb," amongst other powers and licenses claimed by "believers." As mentioned earlier, whole sections of scripture may be recited from memory or read aloud while casting an intention over a space, with Psalm 23 a popular choice.

For some rootworkers, loudly banging pots and pans throughout the space to clear it of energetic debris, ghosts, and haints is a first go-to, and this is a method probably dating back at least hundreds of years, given its sheer simplicity. A root doctor initiated into a family lineage of hoodoo once relayed to me that she found this method to be effective at causing tricked objects hidden within a home to surface and reveal themselves for further removal and nullification.

HO'OPONOPONO

In the case of sound as a healing device, Ho'oponopono is a Hawaiian healing technique that, in four short phrases expresses acknowledgment, apology, gratitude, and love toward reconciling energies that are out of alignment—

"I love you, I'm sorry, forgive me, and thank you." Though the intricacies of this modality are deeper than can be explored here, the spirit of it as expressed in these four phrases can completely shift the energy in a space.

A Hoʻoponopono-inspired clearing can be performed by speaking aloud these phrases with great sincerity to an entire space or a particularly trauma-filled area. Truly convey the meaning of each to the space and to the energies within it that are out of alignment with one another or with their true essence due to trauma or neglect. To me, this looks like acknowledging presence, sentience, and innate sanctity; apologizing on behalf of whoever or whatever caused it to forget its true nature; asking for forgiveness from a place of oneness with all things, yearning for what is out of touch with its true nature to remember what it is and rejoin the Great Fabric; and thanking it—not for changing at a personal request—but for its true essence.

A loft apartment in Brooklyn is being emptied to make way for new tenants, and the owner wants to cut the energetic ties the tenants have with it. It is large—nearly 2,000 square feet—and cluttered with the disarray that comes with packing, making it less than optimal for floor washing. Time is of the essence.

The words associated with the practice of Ho'oponopono are loudly chanted and sung to clear the space and reconcile it to a state of well-being while the shaman beats a frame drum to amplify the intention. He moves from room to room, spreading the intention for healing and reconciliation into every crevice, reminding the space what it was built to be and thanking it for its true essence. After about a half hour, divination is performed with the tarot, regarding the energetic health of the space. It is very positive. When an outgoing tenant returns to pick up her things to move them into storage, she remarks on how incredible the space feels compared to her previous visit.

QUESTIONS FOR SHAMANIC JOURNEYING

When we undergo a shamanic journey to our trusted helping spirits, we are putting a single question or request into their hands and remaining open to the answer we are given. Sometimes they will work with us to resolve the issue at hand directly within the journey itself, working energetically to clear a space with one of the elements, or casting a golden net of protection around a particularly vulnerable area in our space. They may advise us to craft medicine using the materia magica explored in this book or using techniques and materials that we've never encountered before. They may even suggest that mundane actions be taken that seem unrelated to the issue at hand. We cultivate trust in our helping spirits and in our own ability to accurately interpret the information received in journey states by taking actions based on our interpretations and seeing what changes.

Learning to craft skillful journey questions is an art in itself, but some questions that may be appropriate for shamanic journeying to clear conditions in a space include:

SHOW me what is at the root of the energetic imbalances in this space.

SHOW me why I feel out of alignment with this environment.

SHOW me what it best serves to resolve here.

WHAT is at the root of our difficulty sleeping in this room?

WHAT is standing in the way of me truly loving this home?

CHAPTER 4

PROTECTION

Imbuing a space with protection helps maintain its energetic vibration, spiritual cleanliness, and vitality. In turn, both the inhabitants of the space and their goals are fortified against distracting and obstructive elements, whether those goals are increasing client revenue or simply getting a good night's sleep. The more fortified the space, the easier to maintain its health and vitality, and some spaces have greater inherent energetic and spiritual protection by virtue of their architecture, placement, and the materials used in building them.

The sciences associated with the inherent protection of a space are ancient, but many have been lost to time. Geomancy, or divination by working with Earth itself, is one such science that survives in both European and North African forms but is most often used to address the day-to-day concerns of human lives rather than architectural layouts and regional infrastructure. When the principles of feng shui are applied before beginning a project in order to optimize the wellness of an edifice, the shape, size, and directional orientation of the building are considered, along with the influence from other buildings and the landscape surrounding it. Contrary to popular belief, there is more to protection than brute force; balance and harmony play a large part as well.

Where such exacting arts and sciences as feng shui were not available to help ensure harmony and protection for individual dwellings and the larger community, complex shamanic rites, regional folk traditions, and orally transmitted familial rites were employed, and they can still be found among surviving indigenous cultures. In ancient Europe, "the practices of observing the phases of the moon and the direction of the wind, and orienting construction in alignment with the cardinal directional points, all reveal that building a house was a religious act rife with consequence," observes Claude Lecouteux in *The Tradition of Household Spirits: Ancestral Lore and Practices.* "This is a fact that can be confirmed by a number of instructions and taboos. In Russia, the master builder had to purify himself before setting to work." Early Russian traditions also held that the house itself was a microcosm of the universe, where "the *izba*, the corner where the icons are kept, is the dawn, the ceiling represents the celestial vault, and the

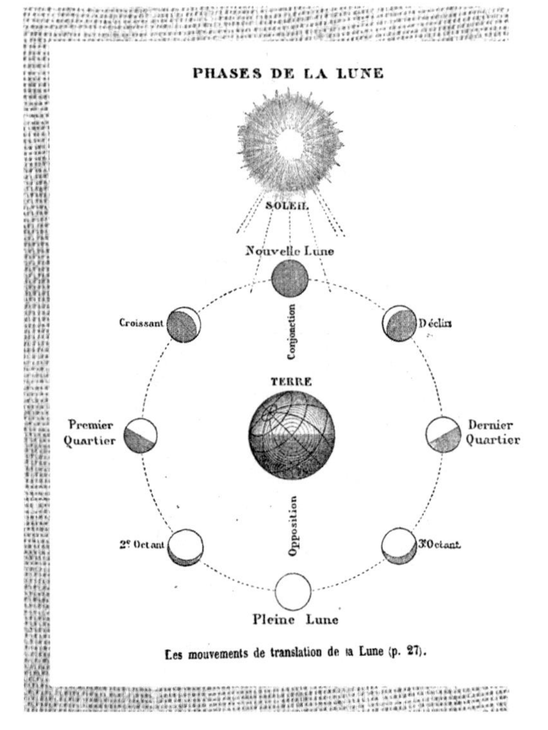

Les mouvements de translation de la Lune (p. 27).

large center beam represents the Milky Way." Among the Dagara of West Africa, a line of ash is used as the foundation of the home, and Ancestral Helping Spirits are called upon to guard it. In Dagara elder Malidoma Some's *Ritual: Power, Healing, and Community*, he recounts an attack against his family by a local witch, resulting in a death. With the home sitting on a continual line of ash, it is widely accepted that the perpetrator—even in animal-spirit form—should not have been able to get in. But a loophole—created by his father not having settled certain ritual obligations—weakened this cornerstone of his family's survival.

> *"Father, why are they putting ash around our house?" He bent down and whispered in my ear, "It is to keep any more evil from coming upon us. Vulture spirits are attracted to deaths like this and try to see if they can cause even more trouble."*

In the contemporary, industrialized world, this way of being with the Earth, cosmology, and its population of spirits has been alternately forgotten and denigrated. Though some of us who build homes and edifices may indeed choose to do so in harmony with nature as best we know, most of us are left to eke out shelter for ourselves in spaces that have been erected without a thought to technologies that work to balance and harmonize. Fortunately, there are ways by which we can greatly augment the protection and fortification of a space to keep us, for the most part, from feeling the need to tear down structures and start all over again.

As a full-time spiritworker, I have indeed at times questioned an architect's sanity or, rather, the sanity of a culture that could produce the messes that we do. In both cases when this happened, I was living in New York City, a region I imagine is even more densely populated with ghosts than with its eight million human inhabitants. Twice have I found myself in a home office environment where my work was under siege by unseen forces. The first time, it did not occur to me how profoundly easy it was for ghosts to wander in and out of spaces, or that altars covered in magic candles, statues, and talismans might be quite the enticing lure for unresolved dead spirits bored of Brooklyn hipster talk who were actively avoiding their very necessary journey to the land of the dead. Since this possibility was not within my worldview, mechanical forms of divination such as the tarot left me wondering why some of my work wasn't resulting in the outcome I expected. It was shamanic journeying that led me to the answer, showing me how the skylight in the first apartment provided the perfect portal for such intrusions, and how almost the entirety of the second apartment was devoid of any protection, due to the way it was constructed in relation to the rest of the building. In spite of the physical architecture, it was as if it had no ceiling at all! In the former space, a simple herbal protection talisman hung in the ceiling's recess did the trick. With the latter space in close proximity to a large cemetery, the situation sometimes seemed hopeless, but a hefty amount of ingenuity, strategy, and spirit negotiations eventually paid off.

There are thousands of folk magic remedies for protecting spaces, some emphasizing the kinds of intruders one might want themselves and their space to be protected from, and others providing more general security and well-being for a space's inhabitants. In the former category, we find approaches not only for protecting against enemies and thieves but for warding off harmful spirits, poverty, and illness. We also find ways of keeping away legal authorities, bill collectors, and landlords—a reminder of folk magic's emphasis on the practical and on empowering marginalized people and communities.

The latter category of protection efforts is more general and likely to be applied at regular intervals as part of a housekeeping routine. Just as one locks the door after entering and leaving the house, solitary practitioners should aim to cultivate habitual rhythms around protective acts that feel potent to them, speak to both their needs and tastes, and can be built upon or amended, as circumstances require. Some kind of protective act should always follow acts of cleansing and removal, and the standard cleansing floor wash is a perfect place to begin our explorations.

FIXING A SPACE

In rootwork, "fixing" something means that one has applied intention to an object, person, or space via magical means. After performing the floor wash discussed in the previous chapter and emptying the scrub bucket of its contents at a crossroads or in the front yard, a new floor wash can be made by adding commercial condition bath crystals, herbal baths, or oils to a bucket of hot water, or an infusion of botanicals in dry or essential oil form that speaks to the qualities you want to bring into your home can be added. The new floor wash is then used to mop inward—that is, up the front steps or hallway and up to the threshold of the front door, paying close attention to the doorway. In a private house, this scrub can be disposed of in the backyard (where hoodoo workings for attracting and drawing are often disposed of). If you live in an apartment, down the toilet will work just fine.

Though exploring every type of hoodoo condition and associated botanicals or commercially available products is beyond the scope of this book, some materials especially lauded for their protective qualities include:

ALFALFA: revered for helping to protect from poverty and hunger

ANGELICA ROOT: also known as Archangel root and Holy Ghost root, associated with both protection and blessing, especially for women and children

BASIL: "Where basil has been, evil cannot be" is a well-known folk saying

DEVIL'S SHOESTRINGS (CRAMP BARK): for "tripping up" evil and harm

HIGH JOHN THE CONQUEROR ROOT: endowing protection is among its many lauded virtues

RUE: beloved throughout the Americas and Southern Europe for its protective and witchcraft-removing qualities

MARJORAM: aids in ensuring peace and tranquility within the home

ROSEMARY: protective of the mind and of a woman's authority within the home

THYME: aids in the protection of finances

Traditional names of commercially available protective products that can be found in condition oils, bath crystals, and incense form include Protection, Fiery Wall of Protection, and Fear Not to Walk Over Evil.

The above act, especially following a cleansing floor wash, may entirely satisfy the solitary practitioner, but many folks choose to take additional measures toward the "fixing" of their space for protection and other conditions they seek to imbue it with.

One such common measure is the dressing, or anointing, of doors and windows—entryways into the house—with hoodoo condition oils for protecting the space and drawing into it that which is desired. This is usually performed in the form of a five-spot pattern—evoking the power of the crossroads—with the practitioner anointing the four corners and center of the surface. A spiritworker may also choose to use one or more condition oils to draw the sign of the cross above these entryways, particularly if partial to Christian symbolism.

Another common measure taken is a fumigation of the space, using commercially crafted condition incense or dried botanicals and resins. Many practitioners opt to cleanse with the elements of Water and Earth and protect and bless with Air and Fire.

Cedar is another plant often found in smudge bundles throughout the U.S., though less commonly than sage. It is strongly protective and grounding, delineating firm borders and boundaries with a scent that ensures safety and comfort.

FRANKINCENSE AND MYRRH

The biblical combination of frankincense and myrrh continues to be highly favored today amongst occultists and ritualists.

HOUSE BLESSING

Use this for imbuing a space with the energy of blessing.

One part each dried thyme, basil, and frankincense resin

Combine ingredients and sprinkle liberally onto a burning charcoal in a censer carried throughout the space.

TALISMANS, AMULETS, AND WARDS

Talismans and other objects that have been crafted to protect a space are often called wards. They are often placed near the threshold of a space, such as above a doorframe or just inside one's home or office. If there is a need for discretion, they can just as easily be hung behind a framed photo or placed beneath a doormat. Creativity is encouraged.

CHRISTIAN CROSS: A symbol of Christianity widely associated with salvation from oppressive forces and the tranquility that comes with faith and devotion, placing a cross within a room or above a door may be seen as a protective act in and of itself—a way of claiming God's protection of a space.

EQUAL-ARMED CROSS: Far older than the Christian cross as a sign of protection, victory, and power is the equal-armed cross. Its near-ubiquity across time and place finds it in as far-flung places as ancient British magic and within the veves of contemporary Haitian Vodou. In hoodoo, it is yet another symbol evoking the power of the crossroads. Crosses may be

crafted out of tree cuttings, with rowan a popular choice in some European strains of folk magic. Animal bones and household iron nails provide additional possibilities, with the former evoking the power of the dead animal and the latter speaking to the strength and power of both metal and industry.

FIXED STATUES: A statue, doll, or other object may be "fixed" by a practitioner to be a protective guardian of a doorway and household through a process of consecrating and enlivening it. This object is often one of the first lines of defense against harmful energies and, like almost all protective charms and magical items, can be considered apotropaic: if it were unintentionally harmed or marred, that would signify a serious warning to the inhabitants. Divination should be performed to ascertain the cause of the damage. If a threat is detected, the spirit of the fixed object may even warn the inhabitants by showing up in a dream.

MIRRORS: Mirrors deflect energy and are often used in hoodoo spells to reverse harm back to its senders. Small mirrors may be placed on windows or doors facing outward to reverse harm intentions. Practitioners may fix the mirrors first by anointing them with Protection, Fiery Wall of Protection, and/or Reversing condition oils, and/or by smoking them in an appropriate incense blend before employing them around the space.

NAZARS: Having entered numerous New World American traditions by way of Middle Eastern and Mediterranean influences, it is most commonly known as "the evil eye," taking its name from the spiritual condition it is meant to ward off. Most often seen as concentric circles of blue, white, and dark blue in the form of a pendant, bead, or within a Hand of Fatima, it can be placed prominently in a household or place of business to deflect the jealousy and envy of visitors that could sap the well-being and abundance of inhabitants and owners.

RUE: Long-associated with protection from all types of maladies, from the evil eye to overt witchcraft, dried rue stalks can be hung above the front door on the inside of one's home, or a small pouch filled with the cut and sifted plant matter can be hung there instead. For some, a living plant—especially if placed in a window—provides hardy protection just by virtue of its presence in the home.

ANCESTRAL OBJECTS: A fireplace mantel with photos of deceased family members, or an altar set aside specifically for their reverence, is considered by some a protective boon, as the spirits of those whose duty in life

involved watching over the practitioner are now encouraged to do so from beyond the grave. The significance of ancestors as protectors can be found in indigenous and folk traditions around the world; libations and food are offered regularly to appease them and encourage them to act on behalf of the living. A weapon, such as a sword or gun that belonged to an ancestor, may also be considered protective in its own right and might even be displayed prominently.

PAPER: Written intentions and/or sections of sacred text may be inscribed onto paper or parchment and placed near the threshold of an apartment or building. In hoodoo, these texts are most often psalms that have come to be associated with the well-being and protection of the home, but the influence of Jewish tradition on African-American magical practices during the twentieth century resulted in many black households proudly displaying mezuzahs on their doorframes for blessing and protection. Likewise, symbols may be drawn and put up, or they may be applied directly to the surface of a wall or door.

DIRT AND DUST: Red brick dust—especially when made from a brick from an old building—is a beloved hoodoo curio for establishing protective boundaries around a space, presumably because it carries within it the memory of being part of a nearly impenetrable wall. It is made by simply crushing a brick to a fine powder, which is then laid across a threshold to keep out enemies and intruders. Similarly, dirt bought and collected from a place that one associates with safety—such as from around a police precinct or church—can be laid out in a similar fashion or sprinkled within a space in a five-spot pattern. Dirt from the grave of a beloved, protective ancestor is traditionally purchased with a nip of alcohol and a few coins, to be applied to a space in the same fashion as above, directly employing them to safeguard an environment. Where the grave of a trusted ancestor is not available, dirt from the grave of a beloved pet dog would suffice, as would that of an honest person employed in law enforcement when alive.

MOJO BAGS

The most popular and well-known hoodoo talisman is the mojo bag—a cloth or leather pouch containing botanicals, minerals, or zoological curios combined to form the body of a spirit ally that has been "conjured" by the maker of the bag and that is fed regularly with liquid and sometimes incense. Though often a personal item carried by an individual, these are made by some practitioners to protect a home, and such a talisman would be kept near the threshold, probably out of sight.

Over a period of two months, the small protective holy stones, energetically tied to a central, grounding earth shrine, that hang on the doorknobs of every door within a home, fall and break. In the chaos of everyday life, the inhabitants forget to replace them, not realizing that only one is left until it is too late. With their primary ward system effectively nullified, the persistent magical attack launched by enemies against the household begins to wreak havoc on romantic relationships and financial endeavors. A series of divinations are performed and household gods are called to assist in a thorough cleansing of the home from top to bottom. The holy stones are replaced with new ones that have been anointed with protection oils and attuned to the central, grounding earth shrine. The boundaries of the home and its inhabitants' lives have been reset, and they prepare to battle from reconsecrated ground.

HOUSE PROTECTION MOJO BAG

This protective talisman can be crafted for any kind of space. You will need:

2 iron nails
1 piece of red string
Protection or Fiery Wall of Protection condition oil
1 shot of whiskey
1 red or white drawstring flannel bag
1 bat nut
1 whole angelica root
Pinch of dirt collected from the four corners of the space
1 small candle

Begin by taking the two iron nails and the piece of red string and tying them together in the form of an equal-armed

cross, anointing it with Protection or Fiery Wall of Protection oil. Dip the thumb or first two fingers of your dominant hand into the shot of whiskey and make the sign of the cross over the newly created ward, saying, "In the name of the Father, the Son, and the Holy Ghost, I consecrate this cross to protect this space from all harm and evil." Place this in the flannel bag and add the bat nut and whole angelica root, thanking each for being strong in their power of protection and asking them to bring this medicine forward into the mojo bag you are crafting. As a link to the space itself, add the dirt collected from the four corners of the space.

Holding the bag in your non-dominant hand and its drawstring in your dominant one, speak into the bag a declaration that the space the bag is meant for is protected from all harm and evil and that its inhabitants remain healthy and well. Give the bag a name you won't forget, such as "Shield" or "Safety," and tell it what its job is. Finish by saying, "By the power of my words and the power of my breath I bring (name of mojo bag) to life." Inhale deeply and exhale the breath of life into the bag, quickly capturing your breath within it by pulling the

drawstring closed and pinching the bag shut. Tie the drawstring around the mouth of the bag tightly, making sure that no air escapes, and tie it off in nine knots.

Welcome the mojo bag into the world by its name and remind it what it is here to do while anointing it in a five-spot pattern with your condition oil, saying, "(Name), as I feed you, so do you work tirelessly to protect this space." Light your candle and speak to its flame, asking it to further enliven your mojo bag. At this point, you may either hold the bag above the flame (high enough that it doesn't catch fire) while chanting your intention until you feel the energy peak, or you may place the bag between both of your hands and bless it in the heat of the flame by passing it above the candle three times. Feed your mojo bag by anointing it again with condition oil in a five-spot pattern, saying, "(Name), as I feed you, so do you work tirelessly to protect this space." Place it out of sight near the entrance of the space it is meant to protect.

A VARIETY OF REMEDIES

Many practices in folk magic are born out of ingenuity and practicality and don't fit neatly in any one specific category. Here are a few more tricks that have been used throughout the ages by conjurers, root doctors, and everyday folk.

BASIL FLOOR SWEEP: Since evil cannot be where basil has been, it may be sprinkled around a home or room and prayerfully swept out the door.

HERBS IN THE KITCHEN: A bowl of alfalfa or basil may be kept in the kitchen—where the family is tended through the preparation of meals—semi-permanently, with the former botanical warding off total poverty and the latter warding off other maladies.

DEVIL'S SHOESTRINGS: A traditional old-time protective act involves nailing nine whole (rather than cut and sifted) Devil's shoestrings across the threshold of the house to trip up any evil that would seek to enter.

TO STOP EVIL FROM ENTERING THE HOME: In Catherine Yronwode's *Hoodoo Herb and Root Magic*, we are told to "mix sandalwood powder with angelica root power and sprinkle the mix across the front of your house, or place a whole angelica root and sandalwood chips in a muslin bag inside the house or near the front door to repel both evil people and evil spirits."

After weeks of trial and error, the root cause of the continued energetic intrusions within a room is found to be that, energetically, it lacks a ceiling, due to the way it was constructed in relation the rest of the building. The practitioner's helping spirits suggest that a remedy akin to a barbed wire fence be installed. Talismans of a different sort than he is used to crafting will be needed, and live cacti placed in high places will help as well. A spirit bottle will need to be crafted to employ the aid of an entity whose primary job will be to defend the space against energetic intrusions. Last, better mundane and spiritual upkeep of the space will be needed: better organization, weekly floor washes, and banishings before starting each workday.

Branches of holly are harvested from a nearby park, and offerings of tobacco are left for the plant in gratitude. Five tiny corked bottles are filled with thorny blackberry leaves, stinging nettles, pins, and needles. These are tied to the holly branches and hung in the uppermost corners of the room as well as its center. In addition, a mirror facing upward is taped to the ceiling and visualized as deflecting any harmful energies that would enter from above. The energy of the room shifts completely.

Finally, the practitioner journeys to his helping spirits for a suitable servitor and is introduced to a mythical creature who is glad to take on the job of protecting the space. The practitioner is shown the items he needs to place in a decorative bottle to make it a proper home for the creature, along with the weekly offerings that it would prefer.

QUESTIONS FOR SHAMANIC JOURNEYING

WHAT is the best action that I can take to ensure this space's protection from harmful influences?

SHOW me a power object that will effectively ward off the deleterious effect the construction across the street will have on our storefront.

WHERE are the energy leaks in this environment?

SHOW me the source of the attacks on this space.

CHAPTER 5

Haunted Spaces and Intrusive Entities

Spirit intrusions are a very real issue in our world and run the gamut from merely annoying to completely debilitating. When entities attach themselves to a person, the experience can be life-altering. Methods and rites of exorcism to clear people and spaces are found not only in traditional cultures but in the world's major religions as well. Indeed, when it comes to removing intrusive energies, discernment about the kind of entity present and about what it has attached itself to can go a long way toward helping ensure efficacy.

The information in the previous chapters on clearing, cleansing, and protection of spaces can go a long way toward making spaces less vulnerable to intrusive entities and is especially effective in helping to remove energetic debris. What sets intrusive entities apart is their sentience and their will. A wide variety of less-than-helpful entities exist, apart from ghosts and demons, though these two are the ones that are generally recognized in Western culture. In all cases, the presence of harmful entities is experienced as discomforting. In cases of severe demonic possession of a space, physical manifestations of an entity's presence may even occur, such as strange smells in the environment. Those within an entity's sphere of influence may also find themselves prone to depression and accidents.

Spirits of all types are all around us. Spirits with malefic intentions are commonly said to be of "lower energetic vibration" in New Age circles. Though all cosmologies have stories about where humans go when they die, not everyone makes it to what I call the land of the dead, where they resolve their lives and make choices about where it best serves them to go next. Lost and wandering spirits of the deceased that have not moved on naturally or that have not been escorted with the aid of a spiritual practitioner may haunt their surviving family members or an object they felt exceptionally attached to during their lives, such as a house, chair, or piece of jewelry. Those who struggled with various addictions may seek out living people who are struggling with those same addictions and try to siphon their energy. Still others may frequent the locales where they worked or otherwise felt comfortable during their lives, or they may simply attach themselves to any passersby for reasons ranging from affinity to lust. The commonly used terms "restless" and "wandering" are highly appropriate for expressing the general temperament and tendencies of most ghosts. People who died in states of severe emotional trauma may become poltergeists in death, causing furniture to move, lights to flicker, and other seemingly impossible phenomena.

Ghosts may also be employed by the living to carry out tasks for good, ill, and all situations in between, creating a more rare but possible reason for ghostly presence within a space. Though in the Western imagination such rites are thought to be limited to religious practices of so-called "foreign" cultures, the intentional involvement of spirits of the dead as sentient forces in magic and spellwork is as much a part of American folk magic as it is of many other magic-practicing traditions around the world. Though ghosts may be employed to influence the thoughts of the living, retrieve information and objects, or fulfill a whole host of other intentions, they may also be sent to magically attack a person and the person's environment, often carrying instructions to inflict harm to either in specific ways.

Intrusion by demonic entities is, gratefully, more rare. The most extreme manifestations are accompanied by the most seemingly physics-bending phenomena. Along with smelling noxious odors, people experiencing these manifestations can endure terrifying apparitions and even insect infestations, among many other possibilities. Less extreme manifestations may include only one of the above elements or none at all.

The important thing to keep in mind is that a space that is carrying significant amounts of unresolved trauma is the perfect nest for the gathering of wayward souls, poltergeists, and demonic entities.

A multifloor bar on Long Island experiences extreme paranormal occurrences including poltergeist manifestation. The owner experiences cyclical loss of income and has been attacked by what he believes to be a demonic force while sleeping in the basement. A multitude of practitioners, he says, have tried to clear the space, including a group of people working with sage and Tibetan singing bowls over a period of days.

I make an offering of tobacco and cornmeal to the spirits of the land, telling them what the issue is and asking them to lend their aid. A shamanic journey is performed, revealing the source of the poltergeist activity to be the spirit of a woman who lived during the late nineteenth century. The aid of Archangel Raphael is requested to assist in this spirit's departure, but to no avail. I am shown the life of the angry spirit, whose wounds tell

the tale of a woman of means widowed by her husband and then abandoned by her children. Raised to remain dependent on men, she had nowhere to turn and died in loneliness and distress. One by one, my helping spirits and I retrieve emotionally young parts and return them to her until she is at peace enough to depart, escorted to the land of the dead by Archangel Raphael.

The basement energy is shown as distinctly demonic, and Archangel Michael is called on to rid the area of it. He is successful, but a presence remains behind the basement walls, exuding a palpable sense of anxiety during a long and dangerous journey. The image is one of hands pressed against the wall from the outside. When asked what it needs, it answers, "Acknowledgment." A tobacco offering is made to the energy, and the energetic imprint instantly dissipates, leaving the distinct impression that this bar may have been a stop on the Underground Railroad.

GHOSTS: RESTLESS SPIRITS OF THE DEAD

In my own practice, I differentiate between ghosts and ancestral helping spirits. The latter consist of spirits who have, after death, traveled "beyond the veil" to what I term the land of the dead. It is there that they review and resolve their life and, if they choose to, return to be of help to their living descendants. But many spirits do not make this journey, only dying and never moving on, due to issues of attachment, addiction, or trauma; these are the spirits of the dead whose presence results in the mischief and pain we see in the lives of many living people and their environments the world over. These are the ones I term ghosts.

More often than not, it is non-ancestral ghosts that wreak energetic havoc on our environments and, hence, are the ones picked up by spiritual security systems healing practitioners set up as part of their energy work clearing spaces.

This isn't to say that ancestral helping spirits with unresolved issues are not a problem in the lives of the living (quite the contrary), but that they are generally less debilitating when it comes to spaces and, as they are family, their presence is also less prone to set off spiritual alarms. As such, removing their energies from our lives is generally done differently from the way non-familial entities are dealt with.

Rites of ghost removal vary in their intensity with efficacy resting on the skill of the practitioner, the level of attachment of the ghost, and the strength of the tool that is being used. It is a common belief that smudging with sage is all that is needed to remove a ghost (this space-clearing advice is seemingly ubiquitous), but I find it less than efficacious because ghosts are more than merely energetic debris or emotional residue and are sentient, conscious entities. Thus, I find that working with other plant spirits and materials especially attuned to making spaces less habitable for ghosts is more effective. When a spirit intrusion is present in a space (rather than attached to a person), the following are good go-tos if you are new to the work of spirit removal or need first-aid remedies. They are listed in order of efficacy. To use, place them in the corners of the space where the ghostly intrusion is—feel free to apply liberally.

CAMPHOR: This can be purchased in small packages of squares. Place one in each corner of the room. If pets are around, the squares can be placed on shelves or, otherwise, higher than the floor. Smearing a bit of campo phenique or Vicks VapoRub (a camphor-based over-the-counter cream) on the walls is just as effective, and a few drops of camphor essential oil can be added to a base such as olive oil for ease of use as well.

RED WINE VINEGAR: Glasses of red wine vinegar can be placed in each corner of the room where a spirit intrusion is present.

CEDAR: Cedar chips or bark shavings can be placed liberally in the corners of a room to lessen its hospitability to ghostly intrusions. Likewise, cedar essential oil can be added to a base such as olive oil and applied to the walls.

PINE: Pine is a plant burned in certain American Indian traditions especially for keeping spirits of the dead at bay. It can be found in stick form, or its resin or needles can be burned in a censer on a charcoal. Frankincense, due to its high vibrational properties, is also excellent for the removal of most types of low-vibration entities. An excellent general exorcism incense for burning on a lit charcoal, therefore, might combine crushed pine and frankincense resins with a few drops of camphor essential oil. Store in a tight-lidded glass container to keep on hand and follow up with acts of protection after use.

DR. E.'S GHOST TRAP

This technique was taught to me by my mentor and friend, the late root doctor Dr. E. of ConjureDoctor.com, to remove a spirit of the dead from a person or space. Take a dark-colored bottle and add approximately half an inch of beer to it, or pour out the contents of a beer bottle until about that much is left. Add a good pinch of mugwort to this, asking it to be potent in its properties, and gently swirl the mixture around to create an intoxicating spirit treat. Then add nine pins or nails and other objects and plant matter with strongly sticky or thorny qualities. Thorns, nettles, and blackberry leaves are all excellent, but you don't need all of them to get the job done. As you add each of your materials, intend that they keep inside the bottle whatever gets caught.

Place the bottle on a dark cloth in the center of the space and have a cork, the original bottle cap, or a tiny bundle of the plant *espanta muerto* ready (usually available at a local botanica). Light a cigar and tell the spirit that is attached to the space that you are sure it must be tired and in need of sustenance, and that such sustenance is available in the bottle. While saying this, blow smoke on the bottle. You are luring the spirit into it.

At some point, you may hear a "whoosh" or "pop"—the sound of the spirit entering the bottle. I usually see the shape of the smoke above the bottle change into a funnel, at which point I know that the spirit has entered. You may choose to quickly use a yes-no form of divination—such as with a pendulum—to ensure that the spirit has been lured into the bottle before corking or capping it. Wrap the dark-colored cloth around the bottle and use a cord to further bind the bundle. Place it outdoors.

Work of blessing and protecting the space should follow this immediately. Then, bring the captured spirit to a cemetery, leave nine pennies at its gate for safe passage, and dig a hole to bury the bundle, saying prayers for the spirit's passage into the land of the dead.

Of course, any practitioners performing this work should have thorough protections in place on themselves and in the space where it is being performed. Calling upon helping spirits to aid in the luring, trapping, and aftermath is strongly advised.

PROTECTING AGAINST INTRUSIVE ENTITIES

Since sentient intrusive beings are of a different energetic frequency from other intrusive energies, it is helpful to create forms of protection that specifically target them. This is especially true in spaces that have already proven themselves vulnerable to spirit intrusions.

European and American folklore hold that ghosts, haints, and other ghouls are easily distracted when presented with a multitude of tiny objects of the same kind. On sight, they are compelled to count them, and will inevitably miscount and be forced to start all over again, until the dawn breaks and they are chased away by the sun, made to disintegrate, or, in the case of hags, die because they did not make it back to their living bodies. Tales abound of folks awakening to see an apparition hunched over in their room feverishly counting mustard seeds that were sprinkled about or poppy seeds that were laid to trip them up and confuse them. For folks who've found their sleep disturbed by such entities, it once was common practice to place a sifter beneath the bed because the holes were distracting to spirits.

Preventative spirit trap bottles can be crafted that mimic the protective "witch bottles" of Europe. Rather than being a decoy for the person creating them, these bottles are meant to trap spirits inside through the use of herbs, minerals, and curios and can be buried on the premises or hung above a door or window—the places that are most vulnerable in the space.

In crafting a spirit trap bottle, you may consider the following ingredients:

NAILS, PINS, AND/OR NEEDLES: for "nailing down" or "pinning" spirits to the bottle

SPANISH MOSS: for entrapping spirits and causing them to be confused

POPPY AND/OR MUSTARD SEEDS: to confuse and distract spirits

SALT, DEVIL'S SHOESTRINGS, RED PEPPER, AND/OR BLACK PEPPER: to protect and repel harm

BROKEN MIRROR SHARDS: to deflect evil

Place your items in a small bottle with intention, telling each what its job is. Breathe into it to enliven it and stopper it with a cork. Any type of bottle works for this, but you may choose to use a cobalt blue–colored one in accordance with bottle tree lore discussed below. If a spirit trap bottle falls, it should be considered a sign that something bigger than it could handle got to it and that it "took a hit." Bury the old one and make a new one immediately. Then perform divination to find out the most efficacious method to remediate the situation and return harmony to the space and its inhabitants. Otherwise, spirit trap bottles should be buried off-grounds and replaced at least once a year to ensure maximum efficacy.

BOTTLE TREES

Throughout the American South, a different kind of spirit trap can be seen in the yards of homes and businesses. Glass bottles hung from trees or placed directly over the branches of coat racks driven into the ground outdoors are said to keep haints, ghosts, and other spiritual foes from interfering in the lives of the living. Most often, the bottles used are cobalt blue in color, and some say that the mere hanging of the bottle causes ghosts to be attracted to and trapped within them, to be disintegrated by the light of the sun in the morning. Others are less specific in their description of how bottle trees work, maintaining that they are simply a cultural phenomenon rooted in African tradition. I myself have seen bottle trees

on display in Haiti, ending my belief that the practice is Black-American in origin.

Bottle trees are only practical if you own or maintain private property—otherwise, you'll have to be OK with the possibility of your work being undone by passersby or local authorities. Gather empty glass bottles (preferably cobalt blue if you want to remain traditional, though any color bottle will work) and wrap wire around the tops of their necks. If you live in an area that is prone to rain or humidity, it's a good idea to use bottles that still have tops, to avoid stagnant water resting in the bottom. Place the tops back on the bottles and wrap the other end of the wires around sturdy tree branches. Another way to make a bottle tree involves a small dead tree or treelike coat rack whose "branches" can fit inside the mouths of your bottles. After ensuring that the structure is safely secured in the ground, place the bottles directly over the branches, ensuring that they are sturdy and stable.

TRICKSTER SPIRITS AND DEMONIC ENTITIES

As previously stated, our world is populated by spirits of all types, and discernment regarding the type of spirit active during an intrusion is the best first

step toward remediating the situation. Just as there are malefic spirits that were once living humans, so are there spirits that were never human but may enjoy frequenting or occupying places where human activity occurs to feed on energies present there. The trickster spirits discussed here are not the archetypal type associated with indigenous myths detailing the adventures of such characters as Coyote or the Norse god Loki. These are energies that might be considered to be more elemental in their nature, though they may be skilled at disguising themselves as more benefic spirits or as simply wayward souls in need of help. A strong relationship with trusted helping spirits goes a long way toward being able to detect these types of energies early on so that they can be banished from one's sphere of influence using the tools discussed throughout this and

previous chapters. If banishment proves ineffective, it may be that the trickster spirit is a deformed or unhealthy local land or nature spirit with more "rights" to the environment than recent human inhabitants. In this case, healing the spirit or working with healthy spirits of the land as mediators are the best options, and this is further discussed in the next chapter.

An American friend moves to a small apartment in Ecuador next to a rainforest and begins having dreams of a wily figure in her space. She finds that she spends more time away from home than is in her nature, and things of hers have gone missing. Divination indicates that there is a presence in her apartment, but it isn't truly malefic—a trickster spirit from the forest has taken a liking to her and has set up residence in her space. Divination also indicates that appeasement and the establishing of boundaries are the best way to move forward; she puts out a small dish of milk, expresses gratitude for the energy's visit, and requests that this be the last. The entity is not to be seen again.

USEFUL FLOOR WASHES

In Communing with the Spirits, *necromancer Martin Coleman shares a number of recipes for floor washes for making a space trickster spirit–free. I've found them helpful against nearly all types of spirit intrusions.*

MATÉ TEA WASH

This floor wash is discouraging to trickster spirits.

1 level teaspoon maté tea
1 cup boiling water

Add tea to water and allow to steep for eight to ten minutes. Add this tea to a three-gallon bucket of clear water.

DOGWOOD BARK TEA WASH

This floor wash is confusing to trickster spirits.

1 level teaspoon ground dogwood bark
1 cup boiling water

Add tea to water and allow to steep for eight to ten minutes. Add this tea to a three-gallon bucket of clear water.

DOGWOOD BARK ALCOHOLIC WASH

This floor wash is both confusing and debilitating to trickster spirits.

ethyl alcohol or vodka
dogwiid shavings

Add a small quantity of ethyl alcohol or vodka to a Mason jar filled with dogwood bark shavings. Cover the jar and allow it to sit in the refrigerator for a week or more. Add the solution to an equal quantity of clear water. Wash the place to be protected from negative spirit influences. Allow the space to dry before using the area. Renew every three weeks or every month, if necessary.

Demonic entity intrusions are the most likely to be recognized immediately, due to the extreme discomfort they may cause in the space or the patterns of arguments, accidents, and severe emotional disturbance that take place there. Though actions can be taken toward removal of demonic spirits, it is the least of all spirit intrusions to be taken lightly. Novices in the spiritual arts should not attempt to remove demonic forces on their own, as the backlash from angering an entity that has not been effectively removed can result in severe accidents, mental and emotional instability, and even death. Trained and experienced clergy or other professionals should be hired to survey and resolve the situation.

In the meantime, if temporary relocation is not possible, daily anointing of the members of the household with protection oils should accompany the wearing of consecrated protective talismans, and prayers for the safety and deliverance of the inhabitants and household should be made frequently to one's Higher Power. These prayers should not be accompanied by lit candles within the space. If the oppressed are not averse to working within the Catholic tradition, prayers to Archangel Saint Michael (traditionally depicted treading upon a demon) and Saint Jude (patron saint of hopeless causes) should be offered fervently daily until the situation is resolved. In lieu of burning candles within the space, candles may be lit within a Catholic church to accompany these petitions.

A shamanic journey is undertaken in a former hospital that has been turned into an apartment building as a demonstration of space-clearing techniques for a local media outlet. The journeyer is shown a window in a room where, when the space was still a hospital, a woman leapt to her death. The scene of the woman running toward and jumping out of the window is repeated again and again. Directly within the trance state, the journeyer calls in familiar plant helping spirits to help clear the energy of trauma that is on continuous replay; as if wiping chalk from a board, sage and rosemary clear the space and are thanked for doing so. Angelica root is called in to bring healing to the deceased woman, now seated in the space. A spirit that often acts as a psychopomp (or guide for the dead) in the journeyer's practice instantly appears and escorts the woman on to the land of the dead.

Suddenly, it is shown that another intrusion is present; a dead woman, who was a witch while she lived, is seen flying around the room. The journeyer visualizes a protective boundary around the assembled group of people and calls forth greater protection from their helping spirits. The witch, who died earlier than the suicidal woman, had spent the past many decades feeding on the trauma that was left behind and is now angry that her sustenance has been removed. The practitioner requests the assistance of the psychopomp to help move this spirit on, as well, but to no avail. Turning to the spirit they began the journey with, they ask for advice and are shown that in order to accomplish the task, they will need a tool that they brought with them for just such a situation as this.

The practitioner exits the journey, again thanking all of the energies and spirits that came to their aid on behalf of the space and immediately sets up, in ordinary reality, a cauldron consecrated to one of their helping spirits who holds power over realms of the dead. The cauldron will act as an effective spirit trap. It will be especially useful for containing the spirit of a dead person who, while alive, cultivated a strong energy body able to retain much power beyond the grave.

The practitioner calls forth the helping spirit the cauldron is consecrated to and demands that, in the powerful helping spirit's name, the spirit of the witch enter the cauldron and be bound to it until verbally released. A gravitational pull into the cast iron cauldron is felt; the top is placed on it and it is wrapped in black cloth. At a crossroads near the building, the practitioner opens the cauldron and releases the spirit into its helping spirit's domain.

QUESTIONS FOR SHAMANIC JOURNEYING

SHOW me the true nature of the energy I am contending with in this environment.

SHOW me what is causing this space to be vulnerable to intrusive energies.

SHOW me the outcome of my making an alliance with the spirit that is present here.

SHOW me a helping spirit that I can work with to effectively move on spirits of the dead. Show me how it best serves me to work with them.

SHOW me the tools I will need to fully remove this energetic intrusion. How can I protect this space from similar energies in the future?

CHAPTER 6

SPIRITS OF PLACE

Within an animistic worldview, everything has a spirit, including the spaces we inhabit for living, working, and playing. Throughout the world since ancient times, acknowledging these beings and offering gratitude to them has been a common practice, helping to ensure their continuing roles as guardians, protectors, and crucial components in the wheel of life. A thorough treatise on the energetic well-being of spaces would be bereft without a discussion of these beings. They are sometimes difficult to categorize, because throughout their own life cycles they may transform due to perceived need and/or interaction with humans. To effectively engage with them at this level of conversation sometimes requires significant discernment, and it can be of tremendous help when all other avenues of remediation have been exhausted, but it may not be necessary for achieving desired results in most situations.

If concepts of house and land spirits prevailed amongst practitioners of the American folk magic traditions of hoodoo or Pennsylvania Dutch powwow, such experiences and associated acts of remediation have not been recorded. My own foray into this level of remediation began while occupying a shared apartment space in the Williamsburg neighborhood of Brooklyn, New York, during which none of my floor washes, exorcisms, or banishing rituals could rid the house of an entity my spirits continually warned did not favor me. I crafted scriptural talismans and placed them in the area of the home where my spirits indicated that the energetic threat against myself and my home was originating, but they had little to no effect. Finally, during an impromptu spirit contact session with a friend, it came to light that the entity was a house spirit—one that was heavily influential in the unconventional aesthetics of the apartment (my roommate at the time being a painter and fashion designer with somewhat erratic tastes) as well as the social dynamics therein. Though spacious, the apartment wasn't exactly hospitable, and I noticed strange occurrences taking place after the weekly classes I taught there. This energy—in the form of a woolly, cloistered, standoffish man—was annoyed by an abundance of social activity. Hovering around the perimeter of the living room, he was, interestingly enough, not indigenous to the space—his locus was an object brought by my landlord roommate from his family home to this apartment, his abode for the past five to ten years. The entity was the culmination of his childhood memories and family traumas, all wound up in one antisocial consciousness. Though highly reluctant to engage in conversation, the entity admitted that he liked tea, and all hope of progress rested on brewing him a cup.

Unlike spirits of the dead, demonic spirits, and other non-local energies, spirits of place and nature spirits can't be captured or "moved on," and one should expect all attempts to do either to be thwarted, whether through smudging, exorcism, spirit traps, angelic assistance, psychopomp work, or any other modality. No matter how distorted such an energy may become, its original purpose is to serve life—it belongs here—so other acts of remediation must be taken if problems arise in the lives of the living.

As in the example above, I tend to see spirits of place in humanoid form. It is debatable whether that is due to their desire to present themselves as such to me or if it is my own filters, but it works for me, and just as the temperament, personality, and health of a human being can be ascertained in part by their looks, mannerisms, and approachability, so am I able to discern the well-being of such spirits by the same.

HOUSE AND BUILDING SPIRITS

"House spirits" and those of other edifices are energies that may consist solely of the materials used in building the structures, or they may consist of these combined with the impressions and experiences left by previous inhabitants. As in the experience recounted about my Williamsburg apartment, the spirits may be solely the amalgamation of emotional energies imbued in either a stationary or mobile object of significance within the space. In *Nature Spirits and What They Say*, editor Wolfgang Weirauch provides the transcription of an interview conducted with Miller, the house spirit of an old watermill in northern Germany, who recounts his origins as an oak tree spirit of a tree in the Bavarian Forest that was cut down in 1306. Installed as a beam in the mill that year, he identifies as the edifice itself and describes his tasks as ensuring that "the house stays a house," checking it from top to bottom every day to make sure that the floors are level, the other beams stay straight, chaos-loving entities are kept at bay, and that relationships between the house and other local spirits remain amiable. In somewhat typical nature spirit fashion, he admits to playing tricks on the humans who inhabit him for the sheer fun of it. In Claude Lecouteux's *The Tradition of Household Spirits: Ancestral Lore and Practices*, he shares that in pagan Germany, when moving from one place to another, it was considered "necessary to invite the domestic spirit of the former house to follow you into the new one, otherwise it will be left weeping in the abandoned home." If forgotten, it could be retrieved later, but if the spirit of the house was a nuisance, two tiles in the form of a roof could be placed where the dismantled house once stood to keep it there.

Both of these examples demonstrate the close acceptance, almost as a member of the family, with which house spirits are seen in cosmologies that include them. In ensuring both the physical and energetic well-being of the humans who inhabit the spaces where they are, where house spirits exist, they tend to be crucial to at least some part of day-to-day activities, even while remaining completely unknown to those moving in and out of their sphere. As in the case of spirits of the land, the lack of human consciousness about such beings means that these relationships are rarely if ever tended, so if severe trauma occurs or the spirit is offended in some way, reconciliation is almost completely impossible and all other avenues of remediation are sought—everything except addressing the root cause.

The presence of a house or building spirit can be discerned by a number of methods, the easiest of which might be divination with a tool that produces yes/no answers, such as a pendulum. Those with cultivated psychic abilities should be able to sense such spirits once they are alerted to the possibility of their existence or if they intentionally seek them out. Still, those with simply a cultivated "felt sense" for the subtle should be able to discern the presence of such energies and converse with them while in a relaxed, meditative state. My preferred method for conversing with, healing, and negotiating with both house and land spirits is through shamanic journeying.

Once the presence of a house or building spirit has been detected in a space, it is generally a good idea to make a simple offering to it at least once a month to remain in good relationship with it and help ensure that it is carrying out its duties. You may, indeed, ask it what those duties are, what it might need to help it accomplish those tasks more easily, and if there are any problems in the space that it suggests tending to that are beyond the realm of its ability to handle. Offerings may be something simple like a glass of water, a cup of coffee, a piece of fruit, a flower, or some other thing that one of the inhabitants tends to enjoy. It should be placed with clear, spoken intention, such as, "This piece of chocolate is for the spirit of this house. May you never hunger. Your work is appreciated. Thank you." Such offerings can be disposed of after a day. However, though they are an excellent step toward sealing a gap in human-spirit relationships in our world, offerings will not remediate a less-than-optimally functioning or debilitated house spirit and the effect it may have on the inhabitants.

An apartment tenant who is highly adept in the work of magic employs multiple forms of space-clearing to great effect, but still feels that something is off. His living room feels inhospitable and inconducive to the social life he wants to cultivate there. A shamanic journey to see the spirits of place who reside with him reveals a shy house spirit who appears as a young, worn-down man wearing a leather jacket. Sensing into the origins, the tenant sees that he is indeed the spirit of the space, but his etheric body includes the chronic financial anxieties experienced by the tenant's landlord, who lives downstairs in the two-floor apartment building. Energy healing is employed, resulting in the house spirit appearing wearing sunglasses in addition to his leather jacket, no longer a sign of insecurity but of charisma and courage.

Energy healing modalities like Ho'oponopono and Reiki are highly effective techniques for healing spirits of place.

I brew tea for myself and the ornery Williamsburg house spirit daily, drinking mine and leaving his on the kitchen counter as an offering, speaking aloud that the cup is for him and expressing gratitude for his willingness to communicate with me. Lying down on my bed with my headphones broadcasting a didgeridoo track into my ears, I journey to the backroads of the apartment we share. Few words are exchanged; instead, with the aid of my helping spirits, I sense into the places inside his etheric body that are holding on to trauma. I see a mentally ill mother's stifling reliance on her son after a painful divorce from her husband, and the energies of confusion during and after family vacations marked by a father's need to outgas repressed sexual tensions. Over a period of days, layer upon layer is uncovered and released through a combination of energy healing techniques, resulting in the space feeling lighter and lighter each day. Eventually, the antisocial spirit exudes a rainbow energy akin to the titular costume worn by the protagonist in stage productions of Joseph and the Amazing Technicolor Dreamcoat. *The spirit's true nature is unrepressed male creativity and expression, and the apartment's erratic decor is no longer in alignment with the space.*

SPIRITS OF THE LAND

"May the road rise up to meet you."
—Irish blessing

In January 2014, the *New York Times* published an article titled "Workers of the World, Faint!" which explored the phenomenon of *neak ta* (guardian land spirits) possessing factory workers throughout Cambodia and demanding better treatment for employees. Made furious due to a lack of care and acknowledgment in an increasingly industrialized landscape, the spirits, the article noted, made additional demands for monthly offerings of appeasement to compensate for having been displaced by factories and other physical features of the businesses the land now housed.

Though written in a somewhat belittling tone that suggested a kind of unconsciously creative proletariat group-mind reasoning behind the occurrences—the article itself yet another example of the dismissive attitude toward the sanctity of these spirits—it offers insight not only into the common desires of land and nature spirits but the intrinsic connection between land and people.

The term "spirits of the land" includes a wide range of entities. There are Great Spirits of the Land encompassing vast regions that we tend to topographically differentiate as parts of continents by such landscapes as the Appalachian and Rocky Mountain ranges. Then there are more regional spirits of the land that help regulate waterways and wildlife over areas as large as the Northeast tri-state area. And more locally, there are the spirits of land right beneath our feet whose influence may extend one or two city blocks or more. Residing within these boundaries—especially in more wild and natural places—are the "nature spirits" of every type imaginable, including the spirits of the rocks, trees, and plants that populate the area.

Spirits of the land maintain a different type of consciousness from that of humans and other beings, whose lifetimes are incredibly short in comparison. To the Great Spirits of the Land encompassing the area of New York City, it has probably been less than our perception of thirty seconds since a bustling, never-sleeping metropolis sprung up there (and they do wish it were more quiet). Their roles vary, from regulation of wildlife to waste management.

Rivers have spirits. Forests have spirits. This is to say nothing of the spirits of cities and towns themselves—the sentient energies that are an amalgamation of our infrastructures, endeavors, dreams, and experiences, and the geographical location into which we've infused them. Even city parks have spirits—some of which are guardian spirits that didn't leave when the city blocks were built around them. Spirits of the land do get fed up with human endeavors and sometimes depart out of grief at being rendered unable to do their jobs.

When the difficulty that a client is facing in a space is due to troubled land spirits that require my direct interaction with them, I am often told or shown how the people who used to live in the locale tended their relationships with the land

in a way that these spirits considered good. Proper offerings were made and rituals of acknowledgment were conducted that honored not only the land itself but the ecological context in which the land existed. Human progress has led to the overdevelopment of land, with no care how living beings might respond to the endeavors, let alone the sentience of the land itself. Thus, my general expectation when encountering a land spirit is that it will be ornery.

Just as there are innumerable accounts of humans finding peace, sanctuary, healing, and "themselves" out in wild natural places, so are there accounts of humans finding themselves in places where they did not belong. Due to our belief in our dominance over nature, we remain puzzled why certain places locally and globally maintain a reputation for harboring natural psychic phenomena dangerous to human beings, such as certain infamous legendary forests and the Bermuda Triangle. The idea that nature exists for nature's sake—and that certain places are claimed by non-human beings—is common knowledge in indigenous cultures. Even when remediating environments, it is important to distinguish between energies that do not serve life and energies that do, but that remain uncomfortable for human beings. Such circumstances are common throughout the natural world, and we may encounter them closer to home than we expect.

Of course, in areas that have been tended well by humans, or in those that bear little of humanity's mark and are friendly toward us, the experience we are offered is supportive and nourishing. On a subtle level, we know that our experience of well-being is because of the land's own vibrant health, wellness, and happiness, but we have forgotten many of the technologies that restore balance and the ways of life that work to ensure this state. In these situations, one of the things we are experiencing is a heightened level of the support of land spirits whose energy never technically stops where our feet meet the earth. The grounding energy that land spirits provide can be felt at a height that extends even above the tops of our heads, creating a density that makes it harder for energetic intrusions to take residence, while providing a sense of support even as we we tend to the work that fills our days.

An apartment building is haunted and is undergoing construction, leading a sensitive tenant to experience significant difficulties in his health and home life. Divination indicates that the spirits of the land are playing a role, and a shaman circumscribes the block on which the building exists, apologizing on behalf of his species to the spirits of the land and making offerings of red wine and cornmeal wherever it feels appropriate. A few minutes later, those present within the first-floor apartment experience a wave of nausea and delirium as the energetic support of the once-dormant and angry land spirits springs upward through the apartment building, clearing out stuck energies and stabilizing the environment.

Spirits of the land tend to respond well to introductions and protocol. When introducing myself to energies I've not yet met, I always make an offering of tobacco and/or cornmeal. I state my name and that I am known as a medicine person among other humans above the topsoil. I state clearly that I seek to be in good relationship with all beings and try to energetically communicate through images how central this tenet is to my life—as a shaman, artist, and activist. I also explain that I am a part of communities that are doing this work and are supportive of it. Your credentials might not be as detailed, but communicating both goodwill and good intention can go a long way toward garnering some amount of trust and help.

In *Earth, Air, Fire & Water: More Techniques of Natural Magic,* author Scott Cunningham offers a "Rite for the Earth" that, in addition to being a beautiful healing ritual, can serve as a helpful starting point toward forming alliances with the spirits

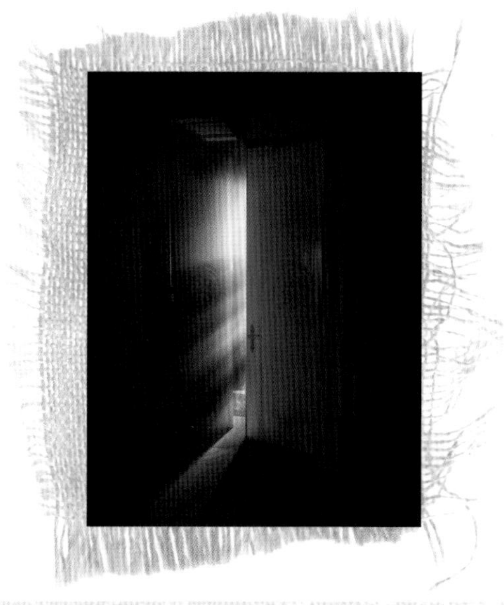

of place in one's area and that may be a helpful addition to your toolbox.

"Light no candles. Burn no incense. Go to the wilds to perform this spell, a place far from the haunts of humans. If this impossible, work this rite in your place of magic. There, have ready a potted plant or four plain stones. (Don't use mined quartz crystals.)

"Sit on the ground. Place your hands onto the dirt beneath you (or onto the plant or stones, if inside). Visualize the Earth from space as a blueish-white sphere of positive, whole, healed energy. Experience it as a living organism.

"When you're ready, say these or similar words:

Pure are the caves and plains;
Pure are the sod and hills;
Pure are the winds and skies;
Pure are the lakes and rills;
Pure are the clouds and rains;
Pure are the woods and trees;
Pure are the valleys deep;
Pure are the bays and seas.

"Renew your visualization, then begin again:

Pure are the birds that fly;
Pure are the hounds and bats;
Pure are the fish and whales;
Pure are the hares and cats;
Pure are the owls and snakes;
Pure are the stags and deer;
Pure are the lizards green;
Pure are all creatures here.

"Once again, renew your visualization. *Feel* the Earth. Block all thoughts of the ravages that our species has subjected to our planet. See the Earth as a healed, whole being. Then say these or whatever words come to mind:

As I receive your energy, now receive mine.

"Pour your personal power out through your palms and into the Earth (or the stones or plant). Gently send it spiraling down into our planet, lending it the strength to survive. Willingly give of yourself for a few moments while visualizing that shining blue planet hanging in the sky.

"After a few moments, end the ritual by lifting your hands and standing. (If performing this rite indoors, place the stones or the potted plant onto the ground outside to allow the Earth to absorb the energy.)

"It is done."

I am interning at an interfaith earth sanctuary in the Allegheny Mountains. As the land is well tended by the sanctuary's wards, it is not only stunningly beautiful but it emits a grounding energy that makes it feel like an abundantly safe place to explore and be true to oneself. Nature spirit sightings are frequent, as the spirits of the land are honored and celebrated here. Within days of setting up my tent in a more barren spot near a river, I begin being followed and threatened by a gigantic bee whenever I enter the area. It takes me a few days to notice that I am being actively told that I am not welcome there. One evening, I arrive at the site and introduce myself aloud, stating that I come in peace and in appreciation of the beauty of this space. I perform Cunningham's "Rite for the Earth," kneeling down on the ground in front of my campsite and giving special attention during my visualizations to the surrounding area. I never see the bee again.

Just as humans absorb energies from below, so does trauma seep into the land when experienced on a mass scale during such events as massacres and war. In such circumstances, though the tools mentioned throughout both this chapter and the previous one might be helpful, it may be even more effective to ask the spirits of the land exactly what they need from the living, in the form of offerings, rituals, prayers, and such, in order to form an alliance, in the hope of working toward healing the trauma that is affecting all parties.

The most commonly discussed concern in the U.S. about land energies potentially affecting homes and businesses is whether construction has taken place on Indian burial grounds. Such places do exist, but as we've explored, these are not the only energies to be concerned about. They are notable, however, because they speak to myriad possibilities that can occur in the interaction between the dead and the land. In circumstances where the buried deceased remain restless and unresolved beneath an edifice, it may be necessary to help them move on, but I've also seen simple appeasements accomplished through the intentional creation of small memorials with stones and pebbles.

The unique possibility that may arise in the case of both Indian burial grounds and other land considerations is born of the relationship between many indigenous peoples and land throughout history. In some tribes, when members die, they go on to become part of the land itself. If they are unhappy and tormented by unresolved conflicts, though they may appear as ghosts, they are actually spirits of the land and will respond as such—that is, minimally—to even the most powerful efforts of exorcism, banishing, and the like. To avoid this, if it is known that indigenous people hold a certain place to be sacred and not fit for building for any reason at all, their word should be final, and other avenues pursued.

In cases of more acute, *tricksterish* nature spirits and elemental energies that may wreak havoc in a home or other environment, actual investigations into their origins, and efforts to appease them with negotiations and simple reconciliatory offerings made at a site of offense, may prove valuable if all avenues of cleansing and protection have been exhausted. Larger and more influential land energies can also become distorted and intrude upon spaces and persons. In

these circumstances, it is important to alert local and nearby spirits of the land regarding what has occurred and to work with them to resolve the issue—returning the land spirit to its proper place. Though centuries of ignorance and abuse have tainted the relationship between humans and the land and caused some nature spirits to act outside their true nature, it is the role of land spirits to uphold, support, and nourish as far as they are able. Humbly but powerfully reminding able spirits of the land of this while engaging in acts of offering, healing, and negotiation can go a long way toward remediation where their help is at least possible, if not an outright necessity.

After many shamanic journeys and cartomantic divinations, it is discovered that the chronic entity problem within a space is not simply a restless dead person, but a land spirit that is also the spirit of an angry indigenous woman. Within a new journey, progress is made using energy healing to heal the wound that was created when her child was taken away from her by the earliest colonists within the area, but she continues to meddle in the affairs of the living in a way that is not conducive to their livelihood. Divination indicates that she will not be placated by offerings or negotiations, and shamanic journeys taken to visit with the spirits of the land from which she originates prove futile, with offerings demanded and supplied over months but no changes seen. Eventually, the practitioner's helping spirits lead him to a much older, larger, foundational land spirit who explains that even the spirits of the land being negotiated with have spent too much time interacting with humans in a way that has distorted them. They are called back by the elder spirit to be renewed into their purpose, and they return as functional land spirits. The practitioner is encouraged to engage the older, foundational land spirit when issues arise.

QUESTIONS FOR SHAMANIC JOURNEYING

SHOW me the influence the spirits of the land are having on this space.

SHOW me the local spirits of the land that it best serves me to form an alliance with toward the protection and well-being of this area. Show me how it best serves me to work with them.

SHOW me why my efforts to clear this space are not meeting my expectations.

SHOW me the household/building spirit/spirits of place present within this environment. Show me how it best serves me to work with them toward the well-being of this space.

CHAPTER 7

Tools for Diagnosing Spaces

It may go without saying, but one of the most helpful tools to consider when diagnosing spaces is lived experience. How do you feel as you move from room to room? Which rooms do you prefer to be in and why? Do you wish certain rooms were more conducive to certain activities than they actually are? How are your sleep patterns? Where do guests spend most of their time? Similarly, for a business environment, which areas of the space get the most foot traffic? Which areas are mostly avoided? If a storefront, do people eagerly window-shop but rarely come in? Ask family members or co-workers for their input as well. Though these questions may seem generic, they invite awareness of the space that you might otherwise overlook and they offer the welcome possibilities of new intentions for a space.

There is yet another seemingly mundane tool that can provide a rich amount of information: *the felt sense.* A term coined by psychotherapist Eugene Gendlin, creator of the process known as Focusing, the felt sense is a subtle bodily awareness of internal or external phenomena. It is generally considered an "unclear," "pre-verbal" "something" available to our consciousness that, in this context, with patient awareness, can give way to minute description.

FELT SENSE EXERCISE

Find a comfortable place in the space you seek to diagnose where you know you won't be disturbed—preferably when the space is empty of other human beings besides yourself. You may choose to lie down if there is little chance of you falling asleep, or you may choose to sit in a comfortable chair.

Close your eyes and take deep breaths that extend all the way into your belly. Begin to relax. Imagine that a silver cord connects from your perineum to the very center of the earth, moving easily downward past the basement of your space—through rock, bones, and varying levels of dense earth. Feel this cord pull taut on your body, and exhale any stress or worry down the cord into the earth, where it explodes and is recycled for use by the universe in other ways. On your next inhale, draw up silver energy from the center of the earth into your body, moving it up the cord as if through a drinking straw. As it pours into your legs and up your torso, imagine that it brings the energy of groundedness and belonging into you, seating you more fully inside your own body. Visualize this energy cascading down your arms and filling you up into your head, pouring out of your mouth and nose to fill your auric space as well. When you feel sufficiently grounded, thank the earth.

Now, bring your awareness to your own body and see if there's a word or phrase that matches your current state of being. The first word that comes to mind might fit perfectly, or you may find it to be only an approximation of how you're *really* feeling. For instance, if you think "relaxed," take that phrase back into your body and see if it fits. You may find that it fits perfectly or that a different, similar word or phrase wants to make itself known. Through patient awareness, you might arrive at "safe," "pleasant," "available," or some other word or phrase that your body simply knows best describes your current state.

Now, bring your awareness to the place where your body and your environment meet. This will most likely be within your aura, from your skin to a few inches outward all around you. With your awareness here, you might see if you can arrive at a word or phrase that best describes your body's felt sense of the room that you're in. These words are not limited to emotional descriptors—your felt sense might be a "pushing in" or a "pulling away," depending on the space's influence on you in this moment. Trust yourself while sensing, again, if there is an even better term that best describes the felt sense you have at this place of awareness. Your body knows.

Now, feel your awareness expand into the space between your body and the walls of the room that you are in (though this exercise can be performed outdoors as well). You may become aware of possibilities that you might be apt to brush off at first notice. Keeping your eyes closed and maintaining a judgment-free, interested curiosity, is there an area of the space that beckons your awareness most? What is your felt sense of that region of the room? If your visualization abilities are strong, you may even realize that you are "seeing" something there through your internal sight. Take note of what you are "seeing" or "knowing." Continue to allow your felt sense to give you information about the area or the energy you're encountering there, and then move on, scanning the room with this bodily awareness.

When finished, write down your experiences and move on to the next room in the space.

With this exercise, even those with minimally cultivated "psychic" abilities should be able to arrive at a relatively in-depth energetic understanding of a space, from the presence of sentient noncorporeal beings to areas where the energy of flow is stagnant or strong. With this in mind, the practitioner might work with a form of divination to arrive at a clearer perspective of the energies encountered and the actions that will help achieve the goals within the space.

ON DIVINATION

Divination is the age-old art of discovering hidden information. There are possibly hundreds of forms of divination, but all of these forms fall into one or more categories. Sortilege, or divination by randomization of objects, encompasses both the well-known tarot card reading as well as Sangoma-style casting of bones. Binary forms of divination—where much of the basis is placed on one or two responses—can be seen in pendulum dowsing and geomancy.

Trance-based forms of divination like shamanic journeying are less popular than mechanical forms of divination, possibly because they take greater skill to wield. In my experience, there is a level of depth that trance-based forms of divination offer that cannot be surpassed; when they are used in conjunction with mechanical forms of divination, the deepest possible clarity emerges. Shamanic journeying became central to my practice precisely because the level of depth it offered was necessary for me to truly understand what was taking place in spaces I was inhabiting, so I cannot overstate its efficacy as an aid, especially when severe and chronic crossed conditions are present within an environment.

I take protocol surrounding divination rather seriously, even though it's not much of a priority in the West. Observing correct protocol ensures that the information I receive is as accurate as possible and as free of distraction and the illusions that energetic intrusions might bring. One way I do this is simply by maintaining a daily practice that consists, in part, of acts of personal spiritual cleansing and protection, connecting with the energies of above and below, inner alignment, and prayer. Before actually performing divination, I call in my helping spirits, or even a specific one. Knowing whom one is conversing with when performing an act of divination is an almost entirely overlooked concept in our culture, with answers received attributed vaguely to "the universe," "the goddess," or the tool itself. This is not to say that receiving direct answers from any

of these is impossible, but that the greater the specificity and direct connection, the higher the divinatory clarity and accuracy. Therefore, it behooves the practitioner first to invoke whomever the questions will be posed to before beginning the act of divination.

Likewise, the cleansing and consecration of divinatory tools is mentioned in many texts and varies based on the materials that your tools are made from. Almost all tools can be cleansed by being smudged with one or more botanicals. Cleansing washes can be made for hardier tools like pendulums, dice, and bone sets. Pendulums typically can be buried in salt overnight, especially as so many commercially produced ones are actual crystals. Cleansing removes any energetic imprints that may burden your tool, and some cleansing routine should be undertaken at least monthly in times of heavy use. Consecration involves energetically imbuing your tool with the intention that it provide you with clear, accurate information and to keep it protected from energetic intrusion. You may consider anointing your tool with one or more hoodoo condition oils, such as Blessing, Clarity, or Fiery Wall of Protection and/or smoking it in the embers of botanicals that lend themselves to similar energies. If the tool is to be used primarily when conversing with a particular spirit, ask for the spirit's blessing upon the tool as well.

Lastly, though divination can provide helpful courses of action, questions beginning with the word "should" are best left out. The sovereignty of the human will is important to maintain while investigating that which is hidden. Impulses to ask "should" questions are best replaced with questions like "Will this action (or series of actions) resolve this problem?" when working with yes/no forms of divination like the pendulum and "What can I expect the outcome to be from this course of action?" when working with tools like the tarot.

PENDULUM DOWSING

Is this space haunted?

Are the spirits of the land here unhappy?

Will the actions I plan to take resolve the energies in this space?

Will XYZ result in this space supporting my business endeavors?

Pendulums are my favorite tool for asking questions that seek a "yes" or "no" answer, and there are some individuals who have tremendous affinity with them. Indeed, whole books have been written on them that go far beyond my knowledge and expertise, so if you feel called to make a pendulum one of your primary tools, research and dig in, as there is a wealth of information out there.

Accurate divination with a pendulum begins with choosing one that suits you. I would suggest picking one out in person. I was taught to do the following:

Take your potential pendulum and hold it by its string or chain between the tips of the thumb and forefinger of your dominant hand directly in front of you. Use the forefinger of your non-dominant hand to nudge the pendulum away from you. As it swings toward you and away, state, "Pendulum, this is my 'yes.'" Use the same forefinger to stop its motion and then to urge it left to right, stating, "Pendulum, this is my 'no.'" Again, bring it to a stop, and then nudge it in clockwise motion, stating, "Pendulum, this is my 'maybe.'" Stop it again and nudge it in a counterclockwise manner, stating, "Pendulum, this is my 'I don't know.'" Stop the pendulum and close your eyes, asking, "Pendulum, are you a good pendulum?" Wait about three to five seconds before opening your eyes. If the answer is no, thank it, place the pendulum back, and find another that calls to you. If the answer is yes, thank it, stop it, close your eyes, and ask, "Pendulum, are you the pendulum for me?" Again, wait about three to five seconds before opening your eyes. If the answer is yes, you've found a new friend. If the answer is no, continue your search.

Once your pendulum has been cleansed and consecrated, you can ask it for the answers to questions the same way you did when choosing it. To take things a step further, you might obtain or create one or more dowsing charts for divining the energetic health or protection of a space, such as a scale from one to ten for the severity of energetic intrusions. With a hand-drawn map of a home or business, your pendulum can help you locate areas of energetic vulnerability and even detect the most efficacious placement of wards and other power objects. Its use is limited only by your imagination. That being said, be sure to cleanse and reconsecrate your pendulum at regular intervals.

THE TAROT

What is the condition of this space?

What is the energy of protection in this space?

What energies are needed in this space to support my endeavors?

What action best serves me to get rid of this intrusive spirit?

The tarot is the most popular tool for divination in the world, falling under the larger category of "cartomancy" or "card divination" under which we find the use of oracle-style decks, Lenormand cards, and standard playing card decks, among others. A tarot deck traditionally consists of seventy-eight cards. Its actual origins are unknown, but that hasn't deterred humanity's immense adoration for it, and many schools of thought, spreads, and styles of decks that have been developed over the ages. Though tarot has been used to provide alternative perspectives through a Western psychoanalytical framework for many decades now, it has been used by healers, root doctors, witches, and shamanic practitioners to diagnose the root causes of spiritual illness for a far longer time.

My own work with the tarot is based more on intuition than rote card memorization or the foundation of formal schools of thought. I was inspired to read the cards this way by Michele Morgan's *A Magical Course in Tarot*, which, at the time of this writing, remains the only book on tarot that I recommend to my own students. As the book you are currently reading is not a primer on learning to read tarot, I suggest that readers familiarize themselves with a deck of their choice. That being said, the interpretations I provide are in no way fixed but are inspired by the imagery found in the Rider-Waite-Smith deck.

Though many have found ways to get accurate "yes" or "no" answers from the tarot, I prefer to leave such questions to binary forms of divination like the pendulum. Tarot lends itself best to questions that seek to understand the overview and nuances of a situation, how others are feeling, what can be expected moving forward, what energies are being brought into the present moment, and what is suggested to create new possibilities.

Though I personally don't use spreads consisting of more than three cards to represent the past, present, and future or to give a general overview of

a situation, I maintain that part of the magic of the tarot is how cards that are near one another interact. For instance, the Seven of Wands, depicting a man on a hill fighting off a horde of staves with one of his own, holds a very different energy when near the Devil card than the Empress. In the former situation, within the context of readings on the environment of a space, the Devil almost always indicates a sentient demonic force; hence, the inhabitants (and perhaps the space itself) are unknowingly battling a strong and malefic entity that is adverse to their well-being. In the latter situation, with the Empress card generally indicating well-being in the home (or the home itself as a place of refuge), if paired with the Seven of Wands it could indicate that a former tenant is still subconsciously or unconsciously attached to the space in such a way that keeps it from being fully energetically available to the new dwellers. When asking for the best steps toward remediation in the latter circumstance, pulling the Eight of Cups or the Three of Swords might indicate the need for cutting the cord between the space and the person—the former by Water, the latter by Air.

When feeling stumped as to the specifics of what is causing disharmony in a space, it is helpful to pull cards while asking more targeted questions regarding specific conditions. Examples include:

WHAT is the condition of this space?

WHAT is the influence the spirits of the land are having on this space?

SHOW me any potential energies of sentient energetic intrusions.

WHAT actions or energies are suggested to increase the well-being of this space?

The following is not meant to be an exhaustive list of the possibilities that diagnosis via tarot can indicate, but a general guideline from my own experience.

THE MAGICIAN: See nearby cards to discern if indicative of supportive or debilitating energies. If the latter, most likely an active spiritual attack on the space or someone in it by a living human being.

THE DEVIL: Sentient harmful force of severely malefic nature.

THE TOWER: Direct energetic attack on the space. Reverse harm and protect immediately.

THREE OF SWORDS: Emotional debris.

FOUR OF SWORDS: A ghost, especially when paired with one of the Court cards or the Moon.

FIVE OF SWORDS: Intentions for space are thwarted. See nearby cards to ascertain source.

SIX OF SWORDS: Leaky. Need for strengthening of space and protection.

SEVEN OF SWORDS: Actively, severely debilitating energies. See nearby cards to ascertain source.

EIGHT OF SWORDS: Stuck and stagnant energies. Severe lack of flow.

NINE OF SWORDS: Severe emotional debris, possibly trauma.

TEN OF SWORDS: A multitude of concerns. A severe situation that is debilitating for inhabitants. Nearly no energies of protection operating effectively.

FIVE OF PENTACLES: Severe energy leaks.

SEVEN OF PENTACLES: Spirits of the land or nature are having an ill effect.

PAGES: Presence of child ghost, trickster energies, and/or disruptive spirits of nature, especially when near the Four of Swords, the Moon, or the Devil.

KNIGHTS: Presence of adolescent ghost or highly active sentient energy.

KINGS, QUEENS: Presence of one or more adult ghosts.

Many tarot readers associate each of the four suits with one of the four natural elements. Keeping these in mind can help generate possible effective remedies for the condition of a space.

SWORDS/AIR: smudging and incense, spoken words and incantations, sound, visualization

WANDS/FIRE: candle magic, dance

CUPS/WATER: floor washes, condition oils, energy healing, fountains

PENTACLES/EARTH: floor sweeps, sachet powders, talismans, wards, work with spirits of the land, work with altars and shrines

SHAMANIC JOURNEYING

Shamanic journeying has been explored throughout this book as an effective modality for diagnosing and remediating spaces.

It can often be helpful to take more than one journey to understand an issue from different possible angles. Journeys can also show us the difference between a space's potential, as opposed to where it is right now, and the steps in between—a visioning process that can provide crucial insight when seeking to completely revamp a space that has chronic energy issues. A journey process like this might look like:

1. *Show me what it would look and feel like if this space (fully met my needs, was fully protected, completely met its potential to serve my goals, and so on).*

2. *Show me what this space looks and feels like now.*

3. *Show me the next action step it best serves me to take toward (goal of question 1).*

4. *Repeat Step 3.*

In this process, what is sought in the first two journeys are three or four words that best describe the experience, including your visuals and felt senses. Considering the symbolic language with which our spirits speak to us in non-ordinary reality, you may indeed be shown things that cannot be installed or done in the space, but instead are meant to convey an idea. For instance, being shown an antique chandelier in your studio apartment in step 1 might be neither practical nor achievable, but it may be a visual cue to remind you about a past memory, or it may simply be the most effective way for your spirits to convey the energies of *elegance* and *attention to detail*. Getting a felt sense of what you are shown in the journey can help you zoom in on what is salient about the symbols you are being provided with.

The helping spirits worked with by humans all have their specialties, and it is widely known that we do not choose our spirits—they choose us, or the spirit world provides us with those who would be most suitable to aid our situation. Though any helping spirit can provide a great deal of information, once we are aware of the spirits that walk with us, it might be wise to consider the types of helping spirits that we are working with in relation to the

questions we need answered. For spotting hard-to-see vulnerable areas in a space, an aviary spirit might come in most handy, especially Owl or Bat, both animals known for seeing or sensing information in the dark. Spirits that have in some way displayed strength and forcefulness might be most efficacious for journeys that are explicitly geared toward resolving sentient intrusion issues.

An office is journeyed about that has chronic energetic intrusion issues. The owner has placed protective talismans in the four directions to ward off both sentient and non-sentient harm, but the intrusions continue. His business partner journeys to see how the office would look and feel if it fully supported their business, how it looks and feels now, and what the next three steps toward achieving the goal posited in the first journey would be. The energetic goal of the actions is of the office feeling like a cathedral—spacious, holy, and protected. They begin to take them.

CHAPTER 8

SAINTS, ANGELS, AND PATRON SPIRITS

In addition to personal and familial helping spirits, nature spirits, spirits of the land, and the spirits of dwellings themselves, there are spirits under whose patronage lies the well-being of homes and businesses. Rather than being tied to the family, land, or dwelling itself, the scope of these spirits tends to be more regional or tradition-based. They are typically categorized as saints or deities, and their influence may be invited in through prayers, offerings, candlelight, the use of objects and recipes associated with them, or even through a permanent shrine dedicated to them. The Ancient Roman goddess Vesta is a perfect example of a spirit associated directly with the home, being patron of the hearth—the place where fire was used to prepare meals and ensure the household's warmth.

SAINTS IN THE HOME

Practitioners of Catholic variations of hoodoo are known to invoke particular saints as visitors to help ensure the efficacy of spiritual work done to cleanse and protect the home, and they sometimes maintain dedicated shrines to them. Through the use of statues, holy cards, art prints, candles with images of saints on them, and other objects, a space may be created for daily, weekly, or monthly communion. Though some may laud the efficacy of simple prayers, there are a multitude of spells associated with the most popular saints and many recipes to call down their influence via floor washes, incense, and oils, with new ones invented every day by the devout and commercial-minded alike. As with all commercial products, it is suggested that you seek out those with actual botanicals and materia magica in them, or those that have been crafted according to traditional cultural standards.

One need not identify as Catholic or Christian to receive the boons that saints offer. Like most spirits of benefic nature who are known worldwide, they tend to respond well to sincere, humble requests and are most influential long-term in the lives of those with whom they have an energetic affinity, regardless of religion. Divination can help foretell what you can expect from working with specific spirits and can even provide a clue as to whom it might best serve you to turn to if the help of a saintly or angelic nature would best serve the situation you're in.

Both the devout and magically astute are keen to ensure that an offering has been provided to a saint following an intercession and in accordance with the saint's preferences. In the case of angels, offerings are not as "necessary"; still, I try to return any favors given with periodic bouquets of flowers and glasses of water set out in gratitude.

ARCHANGEL MICHAEL

Also known as Saint Michael, this being is most often depicted wielding a sword while fiercely subduing a demonic figure. His name means "who is like God?" and he is the ultimate protector angel in the Catholic pantheon, beloved for shielding both persons and spaces from harm of all types, as well as overcoming oppressive spiritual forces that may have already intruded.

A sincere prayer made to Archangel Michael may be all that is needed to invoke his aid in some situations. Oils

bearing his name and used for dressing white or red candles or anointing spaces abound, but Fiery Wall of Protection products as found within the hoodoo tradition are often considered to be well aligned with his force as well. Practitioners who feel especially called upon to have Archangel Michael in their lives may place a statue in their home or workplace somewhere near the entrance or back door as a sign of his protection over the thresholds of the space. Those who work with other spirits might find his force to be so strong that it overpowers the others, and these folks would do well to place his statue outdoors on the property.

SAINT MICHAEL AMPARO

A simple magical method for invoking Saint Michael is through the creation of an *amparo*, a Latin American protection charm. Take a photo of those who are living in the home and place a holy card of Saint Michael over it so that it covers all of the inhabitants. Cut the photo to the same size as the holy card and place a second holy card behind it facing outward, so the image of Saint Michael is "guarding" the inhabitants from all sides. Tie a red string around this like a package and place a white or red tea light on top of it that has been anointed with Archangel Michael or Fiery Wall of Protection oil. In a pinch, olive oil will do. Burn the candle while praying to Saint Michael for the protection of those depicted in the amparo and the space in which they all dwell.

HOME PROTECTION PACKET

Catherine Yronwode's *Hoodoo Herb and Root Magic* describes a New Orleans tradition in which "conjurers used to sew small cloth packets filled with Grains of Paradise and then glue or crochet a holy card of Michael the Archangel to the packets. These were sold in pairs, one to be placed at the front door and one at the back door, for the protection of the home."

ARCHANGEL RAPHAEL

Also known as Saint Raphael, his name means "God heals" in Hebrew, and under his patronage are fishermen, travelers, and those who work in the healing arts. Saint Raphael's story is told in the apocryphal book of Tobit, and two botanicals associated with him are althea and angelica; althea means "healer" in Greek, and angelica means "angel."

Archangel Raphael's relationship to the well-being of spaces is found in his ability to clear an atmosphere of harmful energies—including ghosts and other entities—and imbue the space with healing toward a rebalancing of energies. I often imagine this as an emerald-colored light that he emits throughout the space following the act of cleansing.

As a further use of his healing powers, Raphael can be called upon to aid the healing of deeply wounded spaces and unresolved ghosts too "heavy" to be moved on by simple request. White or green candles can be burned to him, and store-bought or homemade floor washes and incense may be employed to engage magical action alongside his efforts and under his auspices.

SAINT CYPRIAN OF ANTIOCH

Few saints are shrouded in as much mystery as Saint Cyprian of Antioch. The Catholic Church removed him from the official roster in the early twentieth century due to a lack of evidence whether or not he ever lived, but that hasn't curbed the enthusiasm of the practitioners around the world who work with him. With an infamous and often banned European grimoire, *The Book of Saint Cyprian*, bearing his name, he holds an esteemed place in the Americas as found in such traditions as Brazilian Quimbanda, Peruvian shamanism, and Mexican *brujeria*, the latter being the most likely inroads for him to make his way into hoodoo.

This sorcerer-turned-priest is the patron saint of magicians, occultists, and all practitioners of spiritual arts, lauded for his efficacy as both a teacher of magic and commander of spirits, being especially skilled in the work of curse removal, necromancy, and exorcism. He is not as accessible as an archangel or other of the more well-known saints, so it behooves practitioners who want to work with him to begin by building a relationship with him first. Once this has been established, a formidable ally in all works of magic and spirit contact will be had.

Saint Cyprian is the other saint besides Archangel Michael with whom the Latin American amparo is most associated. This protective talisman can be crafted using a holy card or printed image of him, as was done with Saint Michael (see page 97).

Anoint a white candle with an oil for protection or, better yet, one ascribed to him that is available commercially or prepared from a traditional recipe.

Though traditional works associated with him can be performed under his auspices, his enhancement of the practitioner's will, the power of his name, and his force as both a summoner and binder can all be invoked when practicing any work of spirit capture, binding, or exorcism. A popular prayer associated with him involves making the sign of the cross where (+) is written:

Saint Cyprian of Antioch,
I beseech you that those bound by evil spirits and wicked sorceries be unbound. +
I beseech you to shatter all bewitchments and oppressions. +
Save us from the dominion of the wild beasts. +
Saint Cyprian, preserve us from all evil sorceries, spirits,
and malicious arts. +
Guard us in thought, action, and feeling. +
Throw into confusion the wicked ones who seek our lives. +
Confound them with your power. +
Holy Saint Cyprian I beseech you to be our guard and savior. +
By your power may we triumph forever more.
Amen. +

SAINT MARTHA

According to Catholic lore, Saint Martha is the sister of Saint Lazarus and Mary of Bethany (who, by some accounts, is one and the same with Mary Magdalene). Her depiction in the Gospels and apocryphal tales has resulted in her patronage over housekeeping and hospitality. In Catholic hoodoo, a representation of her from France that shows a dragon subdued at her feet invites a mystery as it parallels the most popular image associated with the Mami Wata water spirit pantheon of Africa—a woman dressed in green wielding a snake. In this syncretic form she is known as Saint Martha the Dominator, and her influence extends to one of helping devotees assert control over others—especially women over men (and never the other way around).

Hospitality is providing the service of maintaining control over an environment, so our two Marthas really are one. As an additional testament to her belonging in the folk traditions of spiritual housecleansing, one of her primary symbols is the broom. In *Saint Martha* by Jamie Alexander and S. Aldarnay we find a beautiful rite for consecrating a broom to this saint (see facing page).

PATRON SPIRITS

You are not limited to Catholic saints and angels for the protection and well-being of the home. Personalization is a central aspect of American folk magic practices, so it might be prudent to ask if a spirit with whom you already have an affinity or working relationship is willing to perform the role of guardian over your home or other space. Use whichever way you commune with this spirit to ask, and find

(continued on page 102)

SAINT MARTHA'S BROOM

This ritual broom can be used to clear negativity out of any space. Any household broom can be used, though ideally it should have natural bristles. This sbroom should not be used for mundane sweeping after it has been prepared. You will need:

FOR THE WASH
- *4 camphor tablets*
- *1 bucket of hot water*
- *1 bunch of basil*
- *1 bunch of sage*
- *1 handful of salt*

FOR THE BROOM
- *A new broom*
- *1 bottle of Uncrossing oil*
- *Several feet of green ribbon*

Crush the camphor tablets and put them in the hot water, along with the basil, sage, and salt. Mix the water while reciting Psalm 12.

Set the bucket to one side, and proceed to dress the handle of the broom with the Uncrossing oil. Stand the broom in the bucket, allowing the bristles to absorb the liquid. Leave it to sit overnight. In the morning remove the broom and pour the liquid on your front doorstep. Tie the ribbon around the handle in a criss-cross pattern, from top to bottom. Once the bristles are dry, the broom is ready for use.

To use the broom, simply sweep the house (or any other building with a negative atmosphere) from the back door through to the front door. The broom will pick up any negative or harmful vibrations, energies, or entities and whisk them out of the house. You may wish to repeat Psalm 12 as you sweep for additional potency.

out if there are any particular traits from their lore or personality that would be emphasized in the role. Some spirits are particularly known for shepherding souls of the dead, while others are renowned as guardians over thresholds and archways, making them prime choices for shrines near doors. Still other spirits are known for helping ensure peace and loving vibrations, while some are widely lauded for bringing in customers with high incomes to help ensure prosperity in business. It's good to keep in mind not only what the spirit you're working with is good at and willing to do, but also what the spirit is not willing to do, so that you don't expect what a spirit is not prepared to give or doesn't hold significant influence over.

I once tried to have the spirits whose altars were maintained in the office adjoining my home get involved in the protection of my home, and I was regularly frustrated by their lack of help. When my partner set up a shrine near our front door to renew his relationship with a Greek goddess he once honored fervently, she explained that those spirits were too entrenched in my personal life and client work to take *that* task on too, and stated that she would promptly begin guarding our home with the new shrine installed so prominently.

Should you decide to maintain a permanent shrine, negotiate an arrangement of weekly or monthly offerings. At the allotted time, in addition to refreshing and tidying the space you've provided for the spirit, you might bring them candles, flowers, candy, beverages, foods, and/or other offerings you know they favor. If no traditional recipes exist that are to your liking, you might ask the spirit for a recipe for a floor wash or incense to use that will further invite their presence into the space at regular intervals. Keep in mind the temperament of the spirit you choose and how much space you allot for them as they will almost undoubtedly have some influence on the timbre of the space, most notably if it is in the living room or other highly public environment. The spirit invited might prefer not to have certain activities take place there (or may encourage others altogether!). That being said, even if it's not near the front door of the house, the kitchen remains a traditional place for the keeping and reverence of household spirits due to its centrality to the function of the household.

GLOSSARY

ANIMISM: an umbrella term for all worldviews that posit the existence of a sentient energetic force behind all physical reality and that the universe is made up of many sentient non-human beings, For example, the understanding that structures as big as mountains or as small as stones each have their own unique soul, intelligence, and perspective.

APOTROPAIC: a type of magic or charm, talisman, or statue with the function of turning away harm or evil. When these items are marred or harmed, or if they change in any way, this is often interpreted as a warning to the user. For example, the use of evil eye charms to ward off the evil eye.

CONDITION: a term in hoodoo describing the energetic influences affecting a person or environment. For example, referring to someone who is in some way out of balance spiritually as being *under crossed conditions*. Additionally, the word *conditions* can be used when referring to products created to remediate or engender certain spiritual or energetic circumstances, such as a *condition oil*. Examples of condition oils might include—*Fast Luck*, *Attraction*, *Crown of Success* or *Crossing*.

COSMOLOGY: a cultural map of the universe that explains its origin, evolution, and order and arrangement, usually including stories about non-corporeal sentient beings and their role in relation to the living. The beliefs, principles, and perspectives contained within shared cosmologies instruct how individuals, families, and communities can orient themselves in right relationship to each other and their environment, including both physical and spiritual components of the environment.

DIVINATION: the use of a tool such as tarot cards or a pendulum to discern unseen or unknown influences, whether mental, physical, emotional, or spiritual, upon a person, place, or event.

DOCTRINE OF SIGNATURES: a concept written about by sixteenth century Swiss-German philosopher Paracelsus and found mirrored in healing traditions across the globe, it posits that every plant has physical characteristics that correspond to traits by which it may be deemed helpful for use by humans.

ENERGY HEALING: encompassing a wide range of modalities and techniques, the direct application, removal, or resolution of subtle energies toward the enhancement of the health and vitality of a person, place, or entity.

FIXED, PREPARED: in hoodoo, or African-American folk magic, the imbuing of an object such as a statue, candle, or bedpost with magical and/or spiritual intention, often with the use of herbs, oils, breath, prayer, or ritual.

GEOMANCY: a method of divination that interprets markings on the ground or the patterns formed by tossed handfuls of soil, rocks, or sand. Practiced in Africa, Europe, and the Middle East, there are many variations; the most prevalent form of divinatory geomancy involves the interpretation and analysis of a series of sixteen randomly formed figures.

HELPING SPIRITS: non-corporeal entities that a spiritual practitioner has formed an alliance with, including ancestors, animal guides, spirits of nature, saints, angels, and deities.

HOODOO: also known as *conjure* or *rootwork*, the folk magic tradition cultivated by Black Americans in the southern United States from the late seventeenth century onward; includes additional elements drawn from Native American herbalism, Jewish mysticism, and European folk and grimoire traditions.

HO'OPONOPONO: a Hawaiian energy healing modality focused on reconciliation and forgiveness.

INTRUSIVE ENTITIES: also known as *spirit intrusions*, this refers to sentient non-corporeal energies that are overtly harmful or dependent, draining the vitality of a person or place.

LAND SPIRITS: the sentient energies that comprise the earth and the various landforms that exist upon it, including rivers, forests, volcanoes, oceans, caves, and mountains.

OFFERINGS: actions performed for and objects given to sentient forces such as helping spirits, spirits of place, and elemental energies out of gratitude, to seek appeasement, and/or to maintain right relationship.

REMEDIATION: spiritwork or magical acts performed to alleviate crossed conditions, including spells, rituals, and offerings of appeasement.

RIGHT RELATIONSHIP: an energetically harmonious connection between two or more parties, such as a practitioner and a helping spirit, or a group of people and the land upon which they reside, achieved through proper respect and equal energy exchange, given with mutual understanding of each party's role in the cycle of life.

ROOT DOCTOR: a professional hoodoo practitioner who performs works of folk healing and magic for clients.

SHAMANISM: a term originating among Evenki-speaking peoples of North Asia as *saman*, it later shifted to *shaman* when borrowed into Russian. The term was then applied by Western anthropologists to describe the spiritual practices of larger regions and, eventually, was popularized by Romanian historian of religion Mircea Eliade as a term to describe indigenous spiritual practices all over the world that share similar spiritual technologies. Shamanism ultimately encompasses the premise that shamans are intermediaries or messengers between the human and the spirit worlds. Christina Pratt's *An Encyclopedia of Shamanism*: "A shaman is a healer who works in the invisible world through direct contact with 'spirits.' The invisible world contains all aspects of our world that affect us but are invisible to us, including the spiritual, emotional, psychological, mythical, archetypal, and dream worlds. Shamans use an alternate state of consciousness to enter the invisible world to make changes in the energy found there in a way that directly affects specific changes needed here in the physical world. It is this direct contact with 'spirits' through the use of altered states of consciousness and the movement of energy between the worlds that distinguishes the shaman from other practitioners."

SHAMANIC JOURNEY: a sound-induced trance state used by a shamanic practitioner to travel through non-ordinary reality to interact with helping spirits and retrieve information or engage in acts of remediation.

TAROT: the most popular tool used in cartomancy, or sortilege divination by use of cards.

TRICKSTER SPIRIT: a type of entity that disguises itself as a helping spirit or some other energy that is known to the practitioner but is of dubious intent.

VISUALIZATION: the use of the inner senses of sight, hearing, touch, smell, and taste to experience a person, place, object, or event.

RESOURCES

ASSOCIATION OF INDEPENDENT READERS AND ROOTWORKERS
http://www.readersandrootworkers.org
The Association of Independent Readers and Rootworkers (AIRR) is a gathering of professional practitioners of African American folk magic (hoodoo, conjure, rootwork, and more) who provide psychic reading and spiritual root-doctoring services to the public. AIRR promotes quality service and ethical conduct by means of accreditation and evaluation of participants. Unlike commercial online psychic reading services, AIRR is a participant-supported, not-for-profit directory listing that receives no fees or payments for referrals.

CONJUREDOCTOR.COM
http://www.conjuredoctor.com
Founded in 2008 by the late Dr. E., whose skilled and gifted staff maintain his legacy, ConjureDoctor.com is a purveyor of hoodoo condition oils, powders, herb baths, mojo bags, curios, and more.

LAST MASK CENTER FOR SHAMANIC HEALING
Portland, Oregon
http://www.lastmaskcenter.org
Founded in 2009 by initiated shamanic healer Christina Pratt, author of *An Encyclopedia of Shamanism*, the Last Mask Center for Shamanic Healing provides in-person and remote shamanic healing sessions, monthly journeying circles, and regular workshops and classes for building shamanic skills.

LUCKY MOJO CURIO CO.
Forestville, California
http://www.luckymojo.com
Lucky Mojo is the world's most famous purveyor of folk magic supplies, specializing in a wide range of condition products within the hoodoo tradition. Many of their recipes date back to formulas popular in the early and mid-twentieth century and that were sold at that time by mail-order companies that made hoodoo curios. The site hosts a well-organized forum for customers to discuss ideas for using hoodoo products, as well as a tremendous amount of information on hoodoo theory and in practice.

MOUNTAIN ROSE HERBS
Eugene, Oregon
http://www.mountainroseherbs.com
The premier online purveyor of bulk organic herbs, spices, teas, essential oils, and aromatherapy and herbalism equipment and supplies, Mountain Rose Herbs is renowned for their commitment to quality, as well as their nuanced selection.

OCCUPY YOUR HEART
http://www.occupy-your-heart.com
Founded by shamanic practitioner Langston Kahn, Occupy Your Heart provides in-person and remote shamanic healing and emotional clearing sessions, the latter a body-based energy healing practice that allows past trauma, stuck patterns, outmoded beliefs and stories and addictions to be tracked to their root and transformed there. Kahn teaches workshops and classes in person and online and holds initiations in both African diasporic and European spiritual traditions.

WHY SHAMANISM NOW
http://www.whyshamanism.com
A weekly podcast from host and shamanic healer Christina Pratt focused on the importance of shamanic principles and practices for the Western world. Airing since 2009, *Why Shamanism Now* has consistently provided some of the best conversations on shamanism in contemporary times, including in-depth interviews with spiritual practitioners from across the globe.

BIBLIOGRAPHY

Ali, ConjureMan. *Saint Cyprian*. Yorkshire, UK: Hadean Press, 2011.

Alexander, Jamie and S. Aldarnay. *Saint Martha*. Yorkshire, UK: Hadean Press, 2012.

Armand, Khi. *Deliverance! Hoodoo Spells of Uncrossing, Healing, and Protection*. California: Missionary Independent Spiritual Church, 2015.

Bird, Stephanie Rose. *Sticks, Stones, Roots & Bones: Hoodoo, Mojo & Conjuring with Herbs*. Minnesota: Llewellyn, 2004.

Coleman, Martin. *Communing with the Spirits*. Maine: Weiser, 1997.

Cruden, Loren. *Medicine Grove*. Vermont: Destiny Books, 1997.

Cunningham, Scott. *Earth, Air, Fire & Water: More Techniques of Natural Magic*. Minnesota: Llewellyn, 2012.

Katz, Debra Lynn. *You Are Psychic*. California: Living Dreams, 2015.

Miller, Jason. *Protection and Reversing Magick (Beyond 101)*. New Jersey: New Page, 2006.

Pratt, Christina. *An Encyclopedia of Shamanism*. New York: Rosen, 2007.

Yronwode, Catherine, ed. *The Black Folder: Personal Communications on the Mastery of Hoodoo*. California: Missionary Independent Spiritual Church, 2013.

Yronwode, Catherine. *Hoodoo Herb and Root Magic*. California: Lucky Mojo Curio Company, 2002.

INDEX

Note: Page numbers in **bold** indicate glossary definitions.

J

K

L

M

N

O

P

R

CREDITS

IMAGE CREDITS

Alamy: © Prisma Archivo: 36

Depositphotos: © Konradbak: 77

iStockphoto: © 2Mmedia: 29; © Sava Alexandru: 59; © Maxim Anisimov: 63; © anyaivanova: 64; © asbe: 11; © Marcela Barsse: 34; © Fernando Alvarez Charro: 54; © Classix: 79; © dirkr: 46; © DRAWbyDAR: 89; © duncan1890: 15, 42; © Grigory Fedyukovich: 80; © Floortje: 48; © fotokon: 68; © Chris Gramly: viii; © humonia: 86; © Ben Hung: 78; © Ian_Redding: 22; © ilbusca: 5, 13, 56; © ivanastar: vi; © Jodi Jacobson: 38; © karandaev: 52; © korhankaracan: 16; © LenaKozlova: 101; © lozan365: 94; © Maica: 60; © Craig McCausland: 32; © MireXa: 97; © Missing35mm: 93; © Muenz: 103; © PeopleImages: 37; © piovesempre: 26; © Renphoto: 20, 72; © Rike_: 40; © sedmak: 98; © small_frog: 23; © Alina Solovyova-Vincent: 62 top; © stockcam: 14; © stockyimages: 2, 82; © VeraPetruk: 90; © Juan Manuel Rios Villanueva: 8; © welcomia: 74; © ZU_09: 47

Shutterstock: © Adamlee01: 51; © andreiuc88: 76; © Anton-Burakov: 49; © Blinka: 85; © Binh Thanh Bui: 35; © enmyo: 50; © Everything: xii; © Marzolino: 18; © MK photograp55: 25; © Grigorii Pisotsckii: 31; © spline_x: 45; © Yanfei Sun: 65; © VERSUSstudio: 19

Wellcome Library, London: 33

Courtesy Wikimedia Foundation: 71; Biso: 99; Blake Archive: 58; Fornax: 6; JLPC: 100; NTNU Vitenskapsmuseet: 17; Fiorella Rendón: 62 bottom; Arkady Zarubin: 92

TEXT CREDITS

Chinese Wash recipe, p 92:
Sticks, Stones, Roots & Bones: Hoodoo, Mojo & Conjuring with Herbs by Stephanie Rose Bird © 2004 Llewellyn Worldwide, Ltd. 2143 Wooddale Drive, Woodbury, MN 55125. All rights reserved, used by permission.

Rite of the Earth spell, p 78:
Earth, Air, Fire & Water: More Techniques of Natural Magic by Scott Cunningham © 2002 Llewellyn Worldwide, Ltd. 2143 Wooddale Drive, Woodbury, MN 55125. All rights reserved, used by permission.

ACKNOWLEDGMENTS

For my helping spirits, who have led me through the fires of experience that would allow me to write this book. Thank you for your constant dedication to my clarity and evolution.

For Langston, my ever-encouraging, deeply wise, courageous-hearted dragon. Thank you for seeing me.

For cat and nagasiva yronwode, Dr. E., ConjureMan Ali, the members of AIRR, and all the folk magic practitioners who shared with me their tips, tricks, and worldview. Thank you for carrying the good work forward.

For Christina Pratt, for her joyful, humble, and powerful imparting of some of the most desperately needed teachings in our world today. I look up to you.

For Kate Zimmermann, for your faith in me and constant encouragement as I worked to manifest this text.

For Hannah Reich, whose expert editing was crucial to this endeavor.

ABOUT THE AUTHOR

Johnathan M. Lewis

KHI ARMAND is a spirit-initiated shamanic healer with additional initiations in a variety of New World traditions. He is the author of *Deliverance! Hoodoo Spells of Uncrossing, Protection, and Healing* (Missionary Independent Spiritual Church) and a frequent blogger. Armand holds an MA from NYU in performance studies and a BA from Hampshire College in ritual anthropology.